Adolf Grünbaum

LONDON MATHEMATICAL SOCIETY STUDENT TEXTS

Managing editor: Professor E.B. Davies, Department of Mathematics,
King's College, Strand, London WC2R 2LS

1 Introduction to combinators and λ-calculus, J.R. HINDLEY & J.P. SELDIN
2 Building models by games, WILFRID HODGES
3 Local fields, J.W.S. CASSELS
4 An introduction to twistor theory, S.A. HUGGETT & K.P. TOD
5 Introduction to general relativity, L. HUGHSTON & K.P. TOD
6 Lectures on stochastic analysis: diffusion theory, DANIEL W. STROOCK
7 The theory of evolution and dynamical systems, J. HOFBAUER & K. SIGMUND
8 Summing and nuclear norms in Banach space theory, G.J.O. JAMESON
9 Automorphisms of surfaces after Nielsen and Thurston, A.CASSON & S. BLEILER
10 Nonstandard analysis and its applications, N.CUTLAND (ed)
11 The geometry of spacetime, G. NABER
12 Undergraduate algebraic geometry, MILES REID
13 An Introduction to Hankel Operators, J.R. PARTINGTON

This book is for my son, Aaron.

London Mathematical Society Student Texts. 11

Spacetime and Singularities

An Introduction

GREGORY L. NABER
Professor of Mathematics,
California State University at Chico

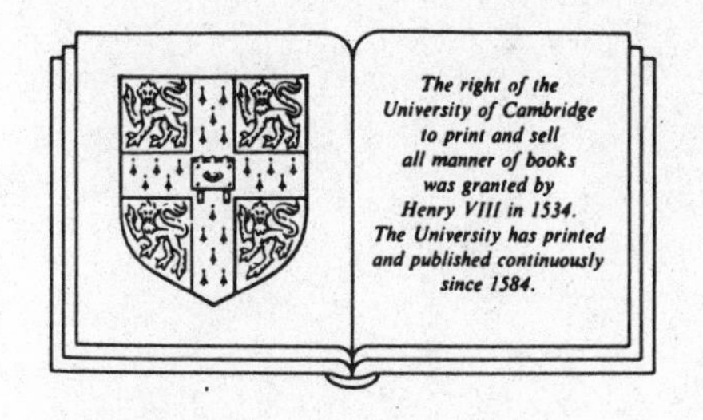

CAMBRIDGE UNIVERSITY PRESS
Cambridge
New York Port Chester Melbourne Sydney

Published by the Press Syndicate of the University of Cambridge
The Pitt Building, Trumpington Street, Cambridge CB2 1RP
40 West 20th Street, New York, NY 10011, USA
10 Stamford Road, Oakleigh, Melbourne 3166, Australia

First published 1988
Reprinted 1990

Printed in Great Britain at the University Press, Cambridge

Library of Congress cataloguing in publication data:

Naber, Gregory L., 1948 -
Spacetime and singularities : an introduction / Gregory L. Naber.
p. cm. -- (London Mathematical Society student texts ; 11)
Bibliography: p.
Includes index.
1. Space and time. 2. General relativity (Physics)
3. Singularities (Mathematics) I. Title. II. Series.
QC173.59.S65N33 1988
530.1'1--dc19

British Library cataloguing in publication data:

Naber, Gregory L. (Gregory Lawrence), *1948* -
Spacetime and singularities
1. Physics. Relativity
I. Title
530.1'1

ISBN 0 521 33327 X hard covers
ISBN 0 521 33612 0 paperback

CONTENTS

PREFACE

In the mid 1960's Stephen Hawking and Roger Penrose each published a series of papers in which were proved the now famous "Singularity Theorems" of general relativity. The theorems are remarkable in that, barring quantum effects, they assert that any "reasonable" model of the universe in which we live must be "singular", that is, contain regions in which the laws of physics as we know them must break down. The theorems of Penrose deal with singularities that arise from gravitational collapse (black holes), while those of Hawking are concerned with the existence of cosmological singularities (the big bang). The intention of this monograph is quite limited and specific: We wish to provide students of mathematics and physics who may have had no previous exposure to relativity and/or differential geometry with a brief, reasonably self-contained and elementary introduction to the ideas required for a rigorous understanding of the simplest of these theorems (essentially that proved by Hawking in [H]). Along the way some of the most fundamental ideas of special and general relativity and modern differential geometry must be introduced and used, but we are quite single-minded in our objective. With the exception of the Problem sets at the end of each chapter there is nothing extraneous to the ultimate goal. In every instance we opt for the most elementary presentation of each idea and we are more than willing to sacrifice generality for clarity. The most obvious example is our decision to consider only manifolds which are explicitly embedded in Euclidean space. One thereby loses much of the elegance of modern differential geometry and even to some extent violates its spirit and that of relativity. Nevertheless, we feel that the geometrical intuition of the beginner is much more easily cultivated within this context. The definitions and arguments very often simplify considerably and so minimize the tendency of the trees to obscure the forest. We can only hope that this brief taste of two complex and beautiful subjects will entice the reader to pursue them to greater depths.

The book is divided into four chapters each of which can be regarded as a reasonable termination point for those whose interests may not extend to a detailed proof of Hawking's Theorem. Chapter 1 is a mathematically sound discussion of the basic kinematical facts of special relativity. The definitions are gotten out of the way in section 1.1 which is followed by a brief discussion of their origins in physics. There follow four sections of linear algebra in which the concepts of section 1.1 are studied in considerable detail and all of the well-known kinematic effects of special relativity (length contraction, time dilation, etc.) are proved as theorems about the geometry of Minkowski spacetime and its group of orthogonal transformations (the Lorentz group).

Chapter 2 contains a *very* brief introduction to some of the ideas of relativistic mechanics. We introduce only those concepts that are crucial to understanding the nature and purpose of the generalization of special relativity discussed in Chapter 3. The emphasis is decidedly mathematical. No detailed experimental procedures are discussed and only a few representative examples are considered.

The background required of the reader in Chapters 1 and 2 is minimal: calculus, a serious course in linear algebra and (for Chapter 2) a nodding acquaintance with some basic Newtonian physics. For Chapter 3 we must ask more. A solid foundation in real analysis (topology of $\mathbb{R}^n$, Implicit and Inverse Function Theorems, etc.) as one might find, for example, in [Sp1] will be indispensible. Here we introduce the basic theory of smooth manifolds and the fundamental ideas of general relativity (the relativistic theory of gravitation). Our goal is to proceed far enough for the reader to understand the *statement* of Hawking's Theorem in section 3.8.

Chapter 4 is rather more technical than those which precede it. It is here that we collect all of the concepts and techniques required to prove Hawking's Theorem. The proof we offer, incidentally, is not Hawking's original proof, but rather one analogous to the proof of Myer's Theorem in Riemannian geometry (see [CE]) and which first appeared in [SW1]. Our decision to restrict attention to cosmological singularities rather than the collapse theorems of Penrose is based primarily on our belief that this proof is more readily made accessible to the audience we have in mind. Sections 4.1 through 4.4 assume the same background as Chapter 3, but will require somewhat more persistence. In the final section we increment our demands upon the reader one last time. This section *sketches* an argument that establishes the existence of length maximizing geodesics in globally hyperbolic spacetimes. In its usual guise this argument makes use of the C^0 topology on a certain set of curves. However, one can avoid asking the reader to learn general topology between sections 4.4 and 4.5 by observing that this C^0 topology is metrizable and, in fact, arises from a rather natural metric. A brief exposure to the theory of metric spaces, available in most point-set topology texts as well as [BG] will suffice for our purposes.

We have no need for general tensor analysis, but do, of course, run across a tensor every now and then. When this happens we go to some lengths to make it clear why the "object" is called a tensor both by reference to its transformation law and its character as a multilinear functional.

A great many more or less routine Exercises are incorporated into the body of the text. These are absolutely essential to the development and must be worked conscientiously. Each chapter is followed by a set of Problems which is intended to expand upon the basic text material. At least read these and work any that strike your fancy. We shall make rather extensive use of the *Einstein summation convention* according to which a repeated index (one subscript and one superscript) indicates a sum over the range of values that index can assume. For example, if a and b are indices that range from 1 to 4, then

$$x^a e_a = \sum_{a=1}^{4} x^a e_a = x^1 e_1 + x^2 e_2 + x^3 e_3 + x^4 e_4,$$

$$\Lambda^a_b x^b = \sum_{b=1}^{4} \Lambda^a_b x^b = \Lambda^a_1 x^1 + \Lambda^a_2 x^2 + \Lambda^a_3 x^3 + \Lambda^a_4 x^4 ,$$

etc.

This text evolved from courses taught at Swarthmore College and the California State University at Chico. Special thanks are due to the faculty and students at these two schools who attended my lectures and contributed much to the completion of this project. To Jim England, Jim Jones and Tom McCready go very special thanks for the support and encouragement they provided. It is also a pleasure to thank Andrea Bowman and Sandy Jones for the patient and cheerful typing of the manuscript. And finally, my thanks to Debora for the artwork, the encouragement, the critical eye and the quiet way in which she relieved me of a thousand small burdens so that I could write instead.

Gregory L. Naber

1987

CHAPTER 1

THE GEOMETRY OF MINKOWSKI SPACETIME

1.1 The Definitions

Minkowski spacetime is a 4-dimensional real vector space $\mathcal{M}$ on which is defined a nondegenerate symmetric bilinear form g of index one; elements of $\mathcal{M}$ will be called *events*. We shall assume that the term "4-dimensional real vector space" is familiar. A *bilinear form* on a vector space V is a map $g : V \times V \to \mathbb{R}$ which is linear in each variable, that is, $g(a_1x_1 + a_2x_2, y) = a_1g(x_1, y) + a_2g(x_2, y)$ and $g(x, a_1y_1 + a_2y_2) = a_1g(x, y_1) + a_2g(x, y_2)$ for all $a_1, a_2 \in \mathbb{R}$ and $x, x_1, x_2, y, y_1, y_2 \in V$. g is *symmetric* if $g(x,y) = g(y,x)$ for all $x,y \in V$ and *nondegenerate* if $g(x,y) = 0$ for all $y \in V$ implies $x = 0$. A nondegenerate, symmetric bilinear form g is generally called an *inner product* and the image of (x,y) under g is often written $x \cdot y$ or $<x,y>$ rather than $g(x,y)$ (provided there is only one inner product under consideration so that no ambiguity will result). The standard example is, of course, the usual inner product on $\mathbb{R}^n$, i.e., if $x = (x^1, \ldots, x^n)$ and $y = (y^1, \ldots, y^n)$, then $g(x,y) = x \cdot y = <x,y> = x^1y^1 + \ldots + x^ny^n$. This particular inner product is *positive definite*, i.e., has the property that $x \neq 0$ implies $g(x,x) > 0$. Not all inner products share this property, however.

Exercise 1.1.1. Define by $g_1(x,y) = x^1y^1 + x^2y^2 + \ldots + x^{n-1}y^{n-1} - x^ny^n$ the map $g_1 : \mathbb{R}^n \times \mathbb{R}^n \to \mathbb{R}$. Show that g_1 is an inner product on $\mathbb{R}^n$, but that there exist nonzero vectors x and y in $\mathbb{R}^n$ with $g_1(x,x) = 0$ and $g_1(y,y) < 0$.

An inner product which is not positive definite is said to be *indefinite*. The following fundamental result is proved in most standard texts in linear algebra, e.g., [H], as well as in [BG].

Theorem 1.1.1. Let V be a finite dimensional real vector space on which is defined an inner product $g : V \times V \to \mathbb{R}$. Then there exists a basis $\{e_1, \ldots, e_n\}$, $n = \dim V$, such that if

$x = x^1 e_1 + \dots + x^n e_n$ and $y = y^1 e_1 + \dots + y^n e_n$ are any two elements of V, then

$$g(x,y) = x^1 y^1 + \dots + x^{n-r} y^{n-r} - x^{n-r+1} y^{n-r+1} - \dots - x^n y^n$$

where r is a non-negative integer (called the *index* of g) that depends only on g.

Thus, Minkowski spacetime $\mathcal{M}$ is a 4-dimensional real vector space with an inner product g (called a *Lorentz inner product*) for which there exists a basis $\{e_1, e_2, e_3, e_4\}$ for $\mathcal{M}$ with the property that if $x = x^1 e_1 + x^2 e_2 + x^3 e_3 + x^4 e_4$ and $y = y^1 e_1 + y^2 e_2 + y^3 e_3 + y^4 e_4$, then

$$g(x,y) = x^1 y^1 + x^2 y^2 + x^3 y^3 - x^4 y^4 .$$

Two vectors x and y in $\mathcal{M}$ are said to be *orthogonal* if $g(x,y) = 0$. An x in $\mathcal{M}$ which satisfies $g(x,x) = \pm 1$ is called a *unit vector*. Any basis for $\mathcal{M}$ such as $\{e_1, e_2, e_3, e_4\}$ is called an *orthonormal basis* because it consists of mutually orthogonal unit vectors; specifically,

$$g(e_a, e_b) = \begin{cases} 1 & \text{if} \quad a = b = 1,2,3 \\ -1 & \text{if} \quad a = b = 4 \\ 0 & \text{if} \quad a \neq b \end{cases} .$$

In the interest of economy we shall introduce a 4×4 matrix η defined by

$$\eta = \begin{bmatrix} 1 & 0 & 0 & 0 \\ 0 & 1 & 0 & 0 \\ 0 & 0 & 1 & 0 \\ 0 & 0 & 0 & -1 \end{bmatrix}$$

whose entries will be denoted either η_{ab} or η^{ab}, the choice in any particular situation being dictated by the requirements of the summation convention (see Preface). Thus,

$$\eta_{ab} = \eta^{ab} = \begin{cases} 1 & \text{if} \quad a = b = 1,2,3 \\ -1 & \text{if} \quad a = b = 4 \\ 0 & \text{if} \quad a \neq b \end{cases} .$$

Thus, for example, we may write $g(e_a, e_b) = \eta_{ab} = \eta^{ab}$ or, using the summation convention, $x = x^a e_a$ and $g(x,y) = \eta_{ab} x^a y^b$, etc.

Given an inner product g on any vector space V we define the *associated quadratic form* $Q: V \to \mathbb{R}$ by $Q(x) = g(x,x) = x \cdot x$ (often denoted x^2). We ask the reader to show that distinct inner products on V cannot give rise to the same quadratic form:

Exercise 1.1.2. Show that if g_1 and g_2 are two inner products on V which satisfy $g_1(x,x) = g_2(x,x)$ for all x in V, then $g_1(x,y) = g_2(x,y)$ for all x and y in V. *Hint:* The map $g_1 - g_2 : V \times V \to \mathbb{R}$ defined by $(g_1 - g_2)(x,y) = g_1(x,y) - g_2(x,y)$ is bilinear and symmetric. Evaluate $(g_1 - g_2)(x+y, x+y)$.

On $\mathcal{M}$ the associated quadratic form is $Q(x) = (x^1)^2 + (x^2)^2 + (x^3)^2 - (x^4)^2$ relative to any orthonormal basis.

If $\{e_1, e_2, e_3, e_4\}$ and $\{\bar{e}_1, \bar{e}_2, \bar{e}_3, \bar{e}_4\}$ are two orthonormal bases for $\mathcal{M}$, then there is a unique linear transformation $L: \mathcal{M} \to \mathcal{M}$ such that $L(e_a) = \bar{e}_a$ for each $a = 1,2,3,4$. Such a map obviously "preserves the inner product" on $\mathcal{M}$, i.e., is of the following type: A linear transformation $L: \mathcal{M} \to \mathcal{M}$ is said to be an *orthogonal transformation* of $\mathcal{M}$ if $Lx \cdot Ly = x \cdot y$ for all $x,y \in \mathcal{M}$. Since the inner product is nondegenerate such a map is necessarily one-to-one and therefore an isomorphism.

Lemma 1.1.2. Let $L: \mathcal{M} \to \mathcal{M}$ be a linear transformation. Then the following are equivalent:

(a) L is an orthogonal transformation.

(b) L preserves the quadratic form of $\mathcal{M}$, i.e., $Q(Lx) = Q(x)$ for all x in $\mathcal{M}$.

(c) L carries any orthonormal basis for $\mathcal{M}$ onto another orthonormal basis for $\mathcal{M}$.

Exercise 1.1.3. Prove Lemma 1.1.2. *Hint*: To prove that (b) implies (a) compute $L(x+y) \cdot L(x+y) - L(x-y) \cdot L(x-y)$.

Now let $L: \mathcal{M} \to \mathcal{M}$ be an orthogonal transformation of $\mathcal{M}$ and $\{e_1, e_2, e_3, e_4\}$ an orthonormal basis for $\mathcal{M}$. By Lemma 1.1.2, $\bar{e}_1 = Le_1, \bar{e}_2 = Le_2, \bar{e}_3 = Le_3$ and $\bar{e}_4 = Le_4$ also form an orthonormal basis for $\mathcal{M}$. In particular, each e_u, $u = 1,2,3,4$, can be expressed as a linear combination of the $\bar{e}_a$:

$$e_u = \Lambda^1_u \bar{e}_1 + \Lambda^2_u \bar{e}_2 + \Lambda^3_u \bar{e}_3 + \Lambda^4_u \bar{e}_4 = \Lambda^a_u \bar{e}_a, \quad u = 1,2,3,4, \tag{1}$$

where the Λ^a_u are constants. Now the orthogonality conditions $e_c \cdot e_d = \eta_{cd}$, $c,d = 1,2,3,4$, can be written

$$\Lambda^1_c\Lambda^1_d + \Lambda^2_c\Lambda^2_d + \Lambda^3_c\Lambda^3_d - \Lambda^4_c\Lambda^4_d = \eta_{cd},\ c,d = 1,2,3,4 \tag{2}$$

or, with the summation convention,

$$\Lambda^a_c\Lambda^b_d\eta_{ab} = \eta_{cd},\ \ c,d = 1,2,3,4. \tag{3}$$

Exercise 1.1.4. Show that (3) is equivalent to

$$\Lambda^a_c\Lambda^b_d\eta^{cd} = \eta^{ab},\ \ a,b = 1,2,3,4. \tag{4}$$

We define the *matrix* $\Lambda = [\Lambda^a_b]_{a,b=1,2,3,4}$ *associated with* L by

$$\Lambda = \begin{bmatrix} \Lambda^1_1 & \Lambda^1_2 & \Lambda^1_3 & \Lambda^1_4 \\ \Lambda^2_1 & \Lambda^2_2 & \Lambda^2_3 & \Lambda^2_4 \\ \Lambda^3_1 & \Lambda^3_2 & \Lambda^3_3 & \Lambda^3_4 \\ \Lambda^4_1 & \Lambda^4_2 & \Lambda^4_3 & \Lambda^4_4 \end{bmatrix}.$$

Observe that Λ is actually the matrix of L^{-1} relative to the basis $\{\bar{e}_a\}$. Heuristically, the conditions (3) assert that "the columns of Λ are orthogonal unit vectors" while (4) makes the same statement about the rows.

We regard the matrix Λ associated with an orthogonal transformation L as a coordinate transformation matrix in the usual way. Specifically, if the event x in $\mathcal{M}$ has coordinates $x = x^1e_1 + x^2e_2 + x^3e_3 + x^4e_4$ relative to $\{e_a\}$, then its coordinates relative to $\{\bar{e}_a\} = \{Le_a\}$ are $x = \bar{x}^1\bar{e}_1 + \bar{x}^2\bar{e}_2 + \bar{x}^3\bar{e}_3 + \bar{x}^4\bar{e}_4$, where

$$\begin{bmatrix} \bar{x}^1 \\ \bar{x}^2 \\ \bar{x}^3 \\ \bar{x}^4 \end{bmatrix} = [\Lambda^a_b] \begin{bmatrix} x^1 \\ x^2 \\ x^3 \\ x^4 \end{bmatrix}. \tag{6}$$

In more detail,

$$\begin{aligned} \bar{x}^1 &= \Lambda^1_1x^1 + \Lambda^1_2x^2 + \Lambda^1_3x^3 + \Lambda^1_4x^4 \\ \bar{x}^2 &= \Lambda^2_1x^1 + \Lambda^2_2x^2 + \Lambda^2_3x^3 + \Lambda^2_4x^4 \\ \bar{x}^3 &= \Lambda^3_1x^1 + \Lambda^3_2x^2 + \Lambda^3_3x^3 + \Lambda^3_4x^4 \\ \bar{x}^4 &= \Lambda^4_1x^1 + \Lambda^4_2x^2 + \Lambda^4_3x^3 + \Lambda^4_4x^4 \end{aligned} \tag{7}$$

which we generally write more concisely as

$$\bar{x}^a = \Lambda^a_{\ b} x^b, \ \ a = 1,2,3,4. \tag{8}$$

Exercise 1.1.5. By performing the indicated matrix multiplications show that (3) (and therefore (4)) is equivalent to

$$\Lambda^t \eta \Lambda = \eta \ , \tag{9}$$

where "t" means "transpose".

Any 4×4 matrix Λ which satisfies (9) is called a *general* (*homogeneous*) *Lorentz transformation*; at times we indulge in a traditional abuse of terminology and refer to the coordinate transformation (7) as a Lorentz transformation.

Exercise 1.1.6. Show that the set of all 4×4 matrices Λ which satisfy (9) forms a group under matrix multiplication, i.e., that it is closed under the formation of products and inverses. This group is called the *general* (*homogeneous*) *Lorentz group* and denoted $\mathcal{L}_{GH}$.

1.2. The Physical Interpretation

Let us begin by unceremoniously laying down the law concerning the physical interpretation of $\mathcal{M}$. This done we will attempt to provide something in the way of motivation. The elements of $\mathcal{M}$ are called "events" and they are to be thought of as actual physical events (occurrences), although we use the term in the idealized sense of a "point-event", i.e., one which has no spatial extension and no duration. One might picture, for example, an instantaneous collision or explosion or an "instant" in the history of some (point) material particle or photon (to be thought of as a "particle of light"). The entire history of a particle (material or photon) is then represented by a continuous sequence of events called the particle's "wordline". An orthonormal basis $\{e_a\}_{a=1}^4$ for $\mathcal{M}$ we think of as defining a "frame of reference" established by some "admissible observer" (essentially an inertial observer of classical physics, but specific requirements will be set out below). If x is an event and $x = x^a e_a$, then (x^1, x^2, x^3) are the three Cartesian coordinates by which the spatial location of x is specified in this frame of reference and x^4 is the time recorded in this frame for the occurrence of x. The frame of reference corresponding to another admissible observer will be identified with another orthonormal basis $\{\bar{e}_a\}_{a=1}^4$ and if $x = \bar{x}^a \bar{e}_a$, then $(\bar{x}^1, \bar{x}^2, \bar{x}^3, \bar{x}^4)$ are again the spatial and time coordinates by which the event x is identified in the new frame. These two sets of coordinates for x are related by the Lorentz transformation $[\Lambda^a_b]$ corresponding to the orthogonal transformation of $\mathcal{M}$ which carries e_a onto $\bar{e}_a$ for $a = 1,2,3,4$.

Thus, $\mathcal{M}$ is a model of the "event world" and how it is seen and described by inertial observers. A great many assumptions about the universe in which we live go into the construction of this model. We shall list a few that lie closest to the surface and, along the way, attempt to clarify the meaning of some of the terms we have used and "justify" the interpretation we have described. Lest you be too easily convinced, however, we shall also pose a few parenthetical questions to ponder.

Assumption #1. Each admissible observer presides over a 3-dimensional, right-handed Cartesian spatial coordinate system based on an agreed unit of length and relative to which photons (light signals) propagate along straight lines in any direction.

(Reasonable enough, but are these coordinate systems expected to cover the entire universe? Can they do that? What are they supposed to be made of? How are spatial coordinates determined in this way related by those actually used by astronomers? What about the small scale world of the atomic nucleus where the uncertainty principle casts doubt upon the very notion of length?)

Incidentally, the expression "presides over" in Assumption #1 is not to be taken too literally. An observer is in no sense ubiquitous. Picture him stationed at the origin of his coordinate system. Any information regarding events that occur at other locations must be communicated to him by means we shall consider shortly.

Assumption #2. Each admissible observer is provided with an ideal standard clock based upon an agreed unit of time with which to provide a quantitative temporal order to the events *on his worldline*.

(What is a "clock" and what does "ideal" mean? Does it mean that the clock can be moved around at will without effecting its "rate"? That it has the same "rate" on Jupiter that it has on earth? Come to think of it what does "its rate" mean? How can we know for even one clock at one location that it is always ticking off "equal time intervals"? (Compare it with another "clock"?))

So now our observer can assign times to the events on his worldline. To do the same for events that are not in his immediate vicinity he will require a system of synchronized clocks throughout his spatial coordinate system (and either assistants or recording devices with each to make the necessary observations and measurements). For the construction of this system of clocks we require some properties of light signals which we collect in our next assumption. First, however, a little experiment: From his location at the origin 0 our observer O emits a light signal at the instant his clock reads t_0. The signal is reflected back to him at a point P and arrives again at 0 at the instant t_1. Assuming there is no delay at P when the signal is bounced back O will calculate the speed of the signal to be dist $(0,P)/\frac{1}{2}(t_1 - t_0)$. This technique for measuring the speed of light we call the *Fizeau procedure* in honor of the gentleman who first carried it out with care (notice that we must bounce the signal back since we do not yet have a set of synchronized clocks).

Assumption #3. For each admissible observer the speed of light *in vacuo* as determined by the Fizeau procedure is independent of when the experiment if performed, the arrangement of the apparatus (i.e., the choice of P) and, moreover, has the same numerical value c (approximately 3.0×10^{10} meters per second) for all such observers.

Here we have the conclusions of two famous experiments first performed by Michelson-Morley and Kennedy-Thorndike. The results may seem odd (why is a photon so unlike an electron whose speed certainly does not have the same numerical value for two observers in relative motion?). Nevertheless, we shall exploit this remarkable property of light signals immediately by asking all of our observers to multiply each of their time readings by c and thereby *measure time in units of distance* (light travel time, e.g., "one meter of time" is the amount of time required by a light signal to travel one meter *in vacuo*). With $x^4, \bar{x}^4, \ldots$ all measured in units of distance all speeds are dimensionless and $c = 1$.

Now we provide each of our admissible observers with a system of synchronized clocks in the following way: at each point P of his spatial coordinate system place a clock identical to that at the origin. At some time x_0^4 at 0 emit a spherical electromagnetic wave (photons in all directions). As the wavefront encounters P set the clock placed there at time $x_0^4 + \text{dist}(0,P)$ and set it ticking, thus synchronized with the clock at the origin. (Why, you may ask, all this fuss about synchronizing clocks? Why not simply synchronize them at the origin and then move them to the desired locations? We shall see.)

At this point each of our admissible observers $O, \bar{O}, \ldots$ has established a frame of reference $\mathcal{S}(x^1,x^2,x^3,x^4)$, $\bar{\mathcal{S}}(\bar{x}^1,\bar{x}^2,\bar{x}^3,\bar{x}^4), \ldots$ relative to which he identifies events (we shall denote the spatial coordinate systems of these frames by $\Sigma(x^1,x^2,x^3)$, $\bar{\Sigma}(\bar{x}^1,\bar{x}^2,\bar{x}^3), \ldots$). How are the $\bar{\mathcal{S}}$-coordinates of an event related to its $\mathcal{S}$-coordinates, i.e., what can be said about the mapping $F: \mathbb{R}^4 \to \mathbb{R}^4$ defined by

$$F(x^1,x^2,x^3,x^4) = (\bar{x}^1,\bar{x}^2,\bar{x}^3,\bar{x}^4) \qquad ? \tag{10}$$

Certainly, it must be one-to-one and onto. Indeed, $F^{-1}: \mathbb{R}^4 \to \mathbb{R}^4$ must be the coordinate transformation from barred to unbarred coordinates. To say more we require a seemingly rather weak "causality assumption".

Assumption #4 (Causality). Any two admissible observers agree on the temporal order of any two events on the worldline of a photon, i.e., if $x = x^a e_a = \bar{x}^a \bar{e}_a$ and $x_0 = x_0^a e_a = \bar{x}_0^a \bar{e}_a$ are two such events, then $x^4 - x_0^4$ and $\bar{x}^4 - \bar{x}_0^4$ have the same sign.

Notice that we do not assume that Δx^4 and $\Delta \bar{x}^4$ are equal, but only that they have the same sign, i.e., that the observers agree as to which of the events occurred first. Thus, F preserves order in the fourth coordinate, at least for events which lie on the worldline of some photon. How are two such events related? Since photons propagate rectilinearly (Assumption #1) with speed 1 (Assumption #2 and our choice of units) two events on the worldline of a photon must have coordinates which satisfy

$$x^i - x_0^i = v^i (x^4 - x_0^4), \qquad i = 1,2,3 \tag{11}$$

for some constants v^1, v^2 and v^3 with $(v^1)^2 + (v^2)^2 + (v^3)^2 = 1$ and consequently

$$(x^1 - x_0^1)^2 + (x^2 - x_0^2)^2 + (x^3 - x_0^3)^2 - (x^4 - x_0^4)^2 = 0 \ . \tag{12}$$

Geometrically, we think of (12) as the equation of a "cone" in $\mathbb{R}^4$ with vertex at $(x_0^1, x_0^2, x_0^3, x_0^4)$ (compare $(z - z_0)^2 = (x - x_0)^2 + (y - y_0)^2$ in $\mathbb{R}^3$) and (11) as a straight line on this cone. But all of this is true in *any* admissible frame so our mapping F must preserve the cone (12). We summarize: the coordinate transformation mapping $F: \mathbb{R}^4 \to \mathbb{R}^4$

A. Carries the cone (12) onto the cone

$$(\bar{x}^1 - \bar{x}_0^1)^2 + (\bar{x}^2 - \bar{x}_0^2)^2 + (\bar{x}^3 - \bar{x}_0^3)^2 - (\bar{x}^4 - \bar{x}_0^4)^2 = 0 \ , \ \text{and} \tag{13}$$

B. satisfies $\bar{x}^4 > \bar{x}_0^4$ whenever $x^4 > x_0^4$ and (12) is satisfied.

Being simply the coordinate transformation from barred to unbarred coordinates the mapping $F^{-1}: \mathbb{R}^4 \to \mathbb{R}^4$ has the obvious analogous properties. In 1964 Zeeman [Z1] called such a mapping F a "causal automorphism" and proved the remarkable fact that any causal automorphism is a composition of the following three basic types:

1. Translations: $\bar{x}^a = x^a + \Lambda^a$, for some constants Λ^a, $a = 1,2,3,4$,
2. Positive scalar multiplications: $\bar{x}^a = kx^a$, for some positive constant k,
3. Linear transformations

$$\bar{x}^a = \Lambda^a_b x^b, \quad a,b = 1,2,3,4, \tag{14}$$

where the matrix $\Lambda = [\Lambda^a_b]_{a,b=1,2,3,4}$ satisfies

$$\Lambda^t \eta \Lambda = \eta \quad \text{and} \tag{15}$$

$$\Lambda^4_4 \geq 1 \ . \tag{16}$$

Notice that it was not even assumed at the outset that $F : \mathbb{R}^4 \to \mathbb{R}^4$ is continuous (much less, linear).

Since two frames of reference related by a mapping of Type 2 differ only by a trivial and unnecessary change of scale we shall banish them from consideration. Since the composition of any number of mappings of Type 1 is obviously again of Type 1 and since we shall prove in section 1.3 that the same is true for mappings of Type 3 we shall restrict our attention to maps of the form

$$\bar{x}^a = \Lambda^a_b x^b + \Lambda^a, \quad a,b = 1,2,3,4 \ . \tag{17}$$

Observe that the constants Λ^a in (17) can be regarded as the barred coordinates of $\mathcal{S}$'s spacetime origin. We should like to assume that the event world is in some sense "homogeneous" so that an observer is free to select any event as his origin (any spatial point as the origin of his Cartesian coordinate system and any instant as his initial instant). Thus, we may request that all admissible observers select the same one.

Assumption #5 (Homogeneity). All admissible observers agree on the spacetime origin, i.e., if $\mathcal{S}(x^1,x^2,x^3,x^4)$ and $\bar{\mathcal{S}}(\bar{x}^1,\bar{x}^2,\bar{x}^3,\bar{x}^4)$ are two admissible frames of reference, then the (necessarily affine) coordinate transformation (17) is, in fact, linear and so assumes the form

$$\bar{x}^a = \Lambda^a_b x^b, \quad a,b = 1,2,3,4, \tag{18}$$

where the matrix $\Lambda = [\Lambda^a_b]_{a,b=1,2,3,4}$ satisfies (15) and (16).

Our next assumption contains, in effect, our definition of "free material particle", the identification of our admissible observers with the inertial observers of Newtonian mechanics, our version of Newton's First Law ("inertial observers exist") and the experimental fact that material particles cannot attain the speed of light relative to an admissible frame.

Assumption #6. Relative to an admissible frame of reference $\mathcal{S}(x^1,x^2,x^3,x^4)$ each point on the worldline of a free material particle has coordinates (x^1,x^2,x^3,x^4) which satisfy

$$x^i - x_0^i = v^i(x^4 - x_0^4), \quad i = 1,2,3, \tag{19}$$

where $(x_0^1,x_0^2,x_0^3,x_0^4)$ are the coordinates of some fixed, but arbitrary point on the worldline and v^1,v^2 and v^3 are constants which satisfy

$$(v^1)^2 + (v^2)^2 + (v^3)^2 < 1 \ . \tag{20}$$

Conversely, any straight line of the form (19) subject to the condition (20) is the worldline of a free material particle.

(Convince yourself that this assumption is consistent with those that precede it, i.e., that all admissible observers agree on which curves are worldlines of free material particles.) We will show somewhat later (Theorem 1.3.3) that our Causality Assumption #4 (whose mathematical expression is (16)) actually forces admissible observers to agree on the temporal order of any two events on the worldline of a free material particle (but *not* on the temporal order of any given pair of events).

With this we hope that the mathematical constructs of section 1.1 are sufficiently well motivated. Nevertheless, we have one more assumption to record. It is the cornerstone upon which the special theory of relativity is built.

Assumption #7 (The Relativity Principle). All admissible frames of reference are completely equivalent for the formulation of the laws of physics.

The Relativity Principle is a powerful tool in building the physics of special relativity, but that is not our task here. For us Assumption #7 is primarily an heuristic principle asserting that there are no "distinguished" admissible observers, i.e., that none can claim to have a privileged view of the universe. In particular, no such observer can claim to be "at rest" while the others are moving; they are all simply in relative motion. We shall see that admissible observers can disagree about some rather startling things (e.g., whether or not two given events are "simultaneous") and the relativity principle will prohibit us from preferring the judgement of one to any of the others.

1.3. The Lorentz Group

A vector x in $\mathcal{M}$ is said to be *spacelike*, *timelike* or *null* (*lightlike*) if $Q(x) > 0$, $Q(x) < 0$ or $Q(x) = 0$ respectively.

Exercise 1.3.1. Use an orthonormal basis for $\mathcal{M}$ to construct a number of examples of each type of vector.

Theorem 1.3.1. Suppose x is timelike and y is either timelike or null and nonzero. Let $\{e_a\}$ be an orthonormal basis for $\mathcal{M}$ with $x = x^a e_a$ and $y = y^a e_a$. Then either (a) $x^4 y^4 > 0$, in which case $x \cdot y < 0$, or (b) $x^4 y^4 < 0$, in which case $x \cdot y > 0$.

Proof. By assumption we have $x \cdot x = x^i x^i - (x^4)^2 < 0$ and $y \cdot y = y^j y^j - (y^4)^2 \le 0$ (summation over $i, j = 1,2,3$) so $(x^4 y^4)^2 > x^i x^i y^j y^j$ and therefore

$$\left| x^4 y^4 \right| > (x^i x^i y^j y^j)^{1/2} . \tag{21}$$

Now, for any real number t, $0 \le (ty^1 + x^1)^2 + (ty^2 + x^2)^2 + (ty^3 + x^3)^2 = (y^j y^j) t^2 + 2(x^i y^i) t + (x^i x^i)$ so, regarded as a quadratic in t, this last expression cannot have distinct real roots and therefore must have discriminant less than or equal to 0, i.e., $4(x^i y^i)^2 - 4(x^i x^i y^j y^j) \le 0$. Thus, $x^i x^i y^j y^j \ge (x^i y^i)^2$ and we find that

$$(x^i x^i y^j y^j)^{1/2} \ge \left| x^i y^i \right| . \tag{22}$$

Combining (21) and (22) yields

$$\left| x^4 y^4 \right| > \left| x^i y^i \right| = \left| x^1 y^1 + x^2 y^2 + x^3 y^3 \right| \tag{23}$$

so, in particular, $x^4 y^4 \ne 0$ and, moreover, $x \cdot y \ne 0$. Suppose, in addition, that $x^4 y^4 > 0$. Then $x^4 y^4 = \left| x^4 y^4 \right| > \left| x^i y^i \right| \ge x^i y^i$ so $x^i y^i - x^4 y^4 < 0$ i.e., $x \cdot y < 0$. On the other hand, if $x^4 y^4 < 0$, then $x \cdot (-y) < 0$ so $x \cdot y > 0$. Q.E.D.

Corollary 1.3.2. If a nonzero vector in $\mathcal{M}$ is orthogonal to a timelike vector, then it must be spacelike.

Another consequence of Theorem 1.3.1 is that, for vectors of the type described there, the product $x^4 y^4$ (which cannot be zero) has the same sign relative to all orthonormal bases for $\mathcal{M}$. If x is timelike and y is either timelike or null and nonzero we shall say the x and y *have the same time orientation* if $x \cdot y < 0$ and *have opposite time orientation* if $x \cdot y > 0$. The analogue of Theorem 1.3.1 for two vectors x and y which are both null, but are not parallel is proved in section 1.5.

Since the orthogonal transformations of $\mathcal{M}$ are isomorphisms and therefore invertible it follows that the matrix Λ associated with such an orthogonal transformation is also invertible (also see (33)). From (9) we then find that $\Lambda^t \eta \Lambda = \eta$ implies $\Lambda^t \eta = \eta \Lambda^{-1}$ so that $\Lambda^{-1} = \eta^{-1} \Lambda^t \eta$ or, since $\eta^{-1} = \eta$,

$$\Lambda^{-1} = \eta \, \Lambda^t \eta \ . \tag{24}$$

To describe the inverse of the coordinate transformation (8) we introduce the following notation: Write each $\bar{e}_u$ as a linear combination of the e_a:

$$\bar{e}_u = \bar{\Lambda}^1_u e_1 + \bar{\Lambda}^2_u e_2 + \bar{\Lambda}^3_u e_3 + \bar{\Lambda}^4_u e_4 = \bar{\Lambda}^a_u e_a \, , \quad u = 1,2,3,4 \ . \tag{25}$$

Then it is clear that the inverse of the coordinate transformation (8) is

$$x^b = \bar{\Lambda}^b_a \bar{x}^a, \quad b = 1,2,3,4 \ . \tag{26}$$

With this notation the orthogonality condition $\bar{e}_a \cdot \bar{e}_b = \eta_{ab}$ can be written

$$\bar{\Lambda}^c_a \bar{\Lambda}^d_b \eta_{cd} = \eta_{ab}, \ a,b = 1,2,3,4 \ . \tag{27}$$

Exercise 1.3.2. Show that (27) is equivalent to

$$\bar{\Lambda}^c_a \bar{\Lambda}^d_b \eta^{ab} = \eta^{cd}, \ c,d = 1,2,3,4 \ . \tag{29}$$

Moreover, from (24) we obtain $[\bar{\Lambda}^b_a] = [\Lambda^a_b]^{-1} = \eta [\Lambda^a_b]^t \, \eta$ and so

$$\begin{bmatrix} \bar{\Lambda}^1_1 & \bar{\Lambda}^1_2 & \bar{\Lambda}^1_3 & \bar{\Lambda}^1_4 \\ \bar{\Lambda}^2_1 & \bar{\Lambda}^2_2 & \bar{\Lambda}^2_3 & \bar{\Lambda}^2_4 \\ \bar{\Lambda}^3_1 & \bar{\Lambda}^3_2 & \bar{\Lambda}^3_3 & \bar{\Lambda}^3_4 \\ \bar{\Lambda}^4_1 & \bar{\Lambda}^4_2 & \bar{\Lambda}^4_3 & \bar{\Lambda}^4_4 \end{bmatrix} = \begin{bmatrix} \Lambda^1_1 & \Lambda^2_1 & \Lambda^3_1 & -\Lambda^4_1 \\ \Lambda^1_2 & \Lambda^2_2 & \Lambda^3_2 & -\Lambda^4_2 \\ \Lambda^1_3 & \Lambda^2_3 & \Lambda^3_3 & -\Lambda^4_3 \\ -\Lambda^1_4 & -\Lambda^2_4 & -\Lambda^3_4 & \Lambda^4_4 \end{bmatrix} \tag{30}$$

Exercise 1.3.3. Show that

$$\bar{\Lambda}^b_a = \eta_{ac} \eta^{bd} \Lambda^c_d \tag{31}$$

and similarly

$$\Lambda_b^a = \eta^{ac}\eta_{bd}\overline{\Lambda}_c^d \ . \tag{32}$$

Note that by taking determinants on both sides of (9) and observing that $\det \eta = -1$ and $\det \Lambda^t = \det \Lambda$, we obtain $(\det \Lambda)^2 = 1$ and so

$$\det \Lambda = \pm 1 \ . \tag{33}$$

Putting $c = d = 4$ in (3) we obtain $(\Lambda_4^4)^2 = 1 + (\Lambda_4^1)^2 + (\Lambda_4^2)^2 + (\Lambda_4^3)^2$ and, in particular, $(\Lambda_4^4)^2 \geq 1$. Consequently,

$$\Lambda_4^4 \geq 1 \quad \text{or} \quad \Lambda_4^4 \leq -1 \ . \tag{34}$$

An element Λ of $\mathcal{L}_{GH}$ is said to be *proper* if $\det \Lambda = 1$ and *improper* if $\det \Lambda = -1$. A Λ in $\mathcal{L}_{GH}$ is *orthochronous* if $\Lambda_4^4 \geq 1$ and *nonorthochronous* if $\Lambda_4^4 \leq -1$.

Improper Lorentz transformations reverse spatial orientation. We now show that nonorthochronous Lorentz transformations reverse time order and therefore violate our basic causality Assumption #4.

Theorem 1.3.3. Let $\Lambda = [\Lambda_b^a]_{a,b=1,2,3,4}$ be a general Lorentz transformation and $\{e_a\}$ an orthonormal basis for $\mathcal{M}$. Then the following are equivalent:

(a) Λ is orthochronous.

(b) Λ preserves the time orientation of all nonzero null vectors, i.e., for every nonzero null vector $x = x^a e_a$ the numbers x^4 and $\bar{x}^4 = \Lambda_b^4 x^b$ have the same sign.

(c) Λ preserves the time orientation of all timelike vectors.

Proof. Let $x = x^a e_a$ be any vector which is either timelike or null and nonzero. By the Schwartz Inequality for $\mathbb{R}^3$ we have

$$(\Lambda_1^4 x^1 + \Lambda_2^4 x^2 + \Lambda_3^4 x^3)^2 \leq ((\Lambda_1^4)^2 + (\Lambda_2^4)^2 + (\Lambda_3^4)^2)((x^1)^2 + (x^2)^2 + (x^3)^2) \ . \tag{35}$$

Now, by (4) with $a = b = 4$ we have

$$(\Lambda_1^4)^2 + (\Lambda_2^4)^2 + (\Lambda_3^4)^2 - (\Lambda_4^4)^2 = \Lambda_c^4 \Lambda_d^4 \eta^{cd} = \eta^{44} = -1 \ . \tag{36}$$

Thus, $(\Lambda_4^4)^2 > (\Lambda_1^4)^2 + (\Lambda_2^4)^2 + (\Lambda_3^4)^2$. Moreover, since x is either timelike or null, $(x^4)^2 \geq (x^1)^2 + (x^2)^2 + (x^3)^2$. Thus, since $x \neq 0$, (35) yields $(\Lambda_1^4 x^1 + \Lambda_2^4 x^2 + \Lambda_3^4 x^3)^2 < (\Lambda_4^4 x^4)^2$, which we may write as

$$(\Lambda^4_1 x^1 + \Lambda^4_2 x^2 + \Lambda^4_3 x^3 - \Lambda^4_4 x^4)\,(\Lambda^4_1 x^1 + \Lambda^4_2 x^2 + \Lambda^4_3 x^3 + \Lambda^4_4 x^4) < 0\ . \tag{37}$$

Define y in $\mathcal{M}$ by $y = \Lambda^4_1 e_1 + \Lambda^4_2 e_2 + \Lambda^4_3 e_3 + \Lambda^4_4 e_4$. By (36), y is timelike. Moreover, (37) can now be written

$$(y \cdot x)\,\bar{x}^4 < 0 \tag{38}$$

Consequently, $y \cdot x$ and $\bar{x}^4$ have opposite signs.

We now show that $\Lambda^4_4 \geq 1$ iff x^4 and $\bar{x}^4$ have the same sign. First suppose $\Lambda^4_4 \geq 1$. If $x^4 > 0$, then $x \cdot y < 0$ (by Theorem 1.3.1) so $\bar{x}^4 > 0$ by (38). Similarly, if $x^4 < 0$, then $x \cdot y > 0$ so $\bar{x}^4 < 0$ by (38). Thus, $\Lambda^4_4 \geq 1$ implies that x^4 and $\bar{x}^4$ have the same sign. In the same way, $\Lambda^4_4 \leq -1$ implies that x^4 and $\bar{x}^4$ have opposite signs. Q.E.D.

Remark. We have actually shown that if Λ is nonorthochronous, then it necessarily reverses the time orientation of *all* timelike and nonzero null vectors.

For these reasons we shall restrict our attention to the set $\mathcal{L}$ of proper, orthochronous Lorentz transformations. From Theorem 1.3.3 it is obvious that the inverse of an orthochronous Lorentz transformation is orthochronous and that the product of two such transformations is again orthochronous. Since the corresponding statements for proper Lorentz transformations are trivial we have proved that $\mathcal{L}$ is closed under the formation of products and inverses, i.e., that $\mathcal{L}$ is a subgroup of $\mathcal{L}_{GH}$. Generally we will refer to $\mathcal{L}$ simply as the *Lorentz group* with the understanding that its elements are proper and orthochronous. The elements of $\mathcal{L}$ are referred to simply as *Lorentz transformations*. Coordinate transformations between admissible frames of reference are all accomplished by elements of $\mathcal{L}$. Henceforth we shall fix some arbitrary orthonormal basis for $\mathcal{M}$ to refer to as the *standard basis* and shall refer to any other orthonormal basis as an *admissible basis* if the coordinate transformation matrix relating it to the standard basis is an element of $\mathcal{L}$.

Remark. One can enlarge the group of coordinate transformations (12) by adding spacetime translations, thereby obtaining the so-called *inhomogeneous Lorentz group* or *Poincaré group*. Physically, this amounts to allowing "admissible" observers to use different spacetime origins.

$\mathcal{L}$ has an important subgroup $\mathcal{R}$ consisting of those $R = [R^a_{\ b}]$ of the form

$$R = \begin{bmatrix} & & & 0 \\ & [R^i_{\ j}] & & 0 \\ & & & 0 \\ 0 & 0 & 0 & 1 \end{bmatrix}$$

where $[R^i_j]_{i,j=1,2,3}$ is a unimodular orthogonal matrix, i.e., satisfies $[R^i_j]^t = [R^i_j]^{-1}$ and $\det[R^i_j] = 1$. Observe that the conditions (3) are obviously satisfied for such an R and, moreover, $R^4_4 = 1$ and $\det R = \det[R^i_j] = 1$ so R is indeed in $\mathcal{L}$. The coordinate transformation associated with R corresponds physically to a rotation of the spatial coordinate axes within a given frame of reference. For this reason $\mathcal{R}$ is called the *rotation subgroup* of $\mathcal{L}$ and its elements are called *rotations* in $\mathcal{L}$.

Lemma 1.3.4. Let $\Lambda = [\Lambda^a_b]_{a,b=1,2,3,4}$ be a proper, orthochronous Lorentz transformation. Then the following are equivalent:

(a) Λ is a rotation.

(b) $\Lambda^1_4 = \Lambda^2_4 = \Lambda^3_4 = 0$.

(c) $\Lambda^4_1 = \Lambda^4_2 = \Lambda^4_3 = 0$.

(d) $\Lambda^4_4 = 1$.

Proof. Set $c = d = 4$ in (4) to obtain

$$(\Lambda^1_4)^2 + (\Lambda^2_4)^2 + (\Lambda^3_4)^2 - (\Lambda^4_4)^2 = -1 \ . \tag{39}$$

Similarly, with $a = b = 4$, (3) becomes

$$(\Lambda^4_1)^2 + (\Lambda^4_2)^2 + (\Lambda^4_3)^2 - (\Lambda^4_4)^2 = -1 \ . \tag{40}$$

The equivalence of (b), (c) and (d) now follows immediately from (39), (40) and the fact that Λ is assumed orthochronous. Since a rotation in Λ obviously satisfies (b), (c) and (d), all that remains is to show that if $[\Lambda^a_b]_{a,b=1,2,3,4}$ satisfies one (and therefore all) of these conditions, then $[\Lambda^i_j]_{i,,j=1,2,3}$ is a unimodular orthogonal matrix.

Exercise 1.3.4. Complete the proof. Q.E.D.

There are 16 parameters in every Lorentz transformation $\Lambda = [\Lambda^a_b]_{a,b=1,2,3,4}$, although only 6 of these are independent by virtue of the relations (3). We now derive simple physical interpretations for each of these parameters. We consider two admissible bases $\{e_a\}$ and $\{\bar{e}_a\}$ and the corresponding frames of reference $\mathcal{S}$ and $\bar{\mathcal{S}}$. Any two events on the worldline of a point which can be interpreted physically as being at rest in $\bar{\mathcal{S}}$ have coordinates in $\bar{\mathcal{S}}$ which satisfy $\Delta\bar{x}^1 = \Delta\bar{x}^2 = \Delta\bar{x}^3 = 0$ and $\Delta\bar{x}^4$ = time separation of the two events as measured in $\bar{\mathcal{S}}$. From (8) we find that the corresponding coordinate differences in $\mathcal{S}$ are

$$\Delta x^b = \overline{\Lambda}^b_a \Delta \bar{x}^a = \overline{\Lambda}^b_4 \Delta \bar{x}^4 \; . \tag{41}$$

From (41) and the fact that, by (3) and (30), $\overline{\Lambda}^4_4$ and $\overline{\Lambda}^4_4$ are non-zero, it follows that the ratios

$$\frac{\Delta x^i}{\Delta x^4} = \frac{\overline{\Lambda}^i_4}{\Lambda^4_4} = -\frac{\Lambda^4_i}{\Lambda^4_4}, \quad i = 1,2,3$$

are constant and independent of the particular point at rest in $\bar{S}$ we choose to examine. Physically, these ratios are interpreted as the components of the ordinary *velocity 3-vector* of $\bar{S}$ relative to S:

$$\vec{u} = u^1 e_1 + u^2 e_2 + u^3 e_3, \quad \text{where } u^i = \frac{\overline{\Lambda}^i_4}{\overline{\Lambda}^4_4} = -\frac{\Lambda^4_i}{\Lambda^4_4}, \quad i = 1,2,3 \tag{42}$$

(we use the term "3-vector" and the familiar vector notation to distinguish such highly observer-dependent spatial vectors, which are not invariant under Lorentz transformations, but are familiar from physics). Similarly, the velocity 3-vector of S relative to $\bar{S}$ is

$$\vec{\bar{u}} = \bar{u}^1 \bar{e}_1 + \bar{u}^2 \bar{e}_2 + \bar{u}^3 \bar{e}_3, \quad \text{where } \bar{u}^i = \frac{\Lambda^i_4}{\Lambda^4_4} = -\frac{\Lambda^4_i}{\Lambda^4_4}, \quad i = 1,2,3. \tag{43}$$

Next observe that

$$\sum_{i=1}^{3} \left(\frac{\Delta x^i}{\Delta x^4} \right)^2 = (\Lambda^4_4)^{-2} \sum_{i=1}^{3} (\Lambda^4_i)^2 = (\Lambda^4_4)^{-2} [(\Lambda^4_4)^2 - 1] \; .$$

Similarly, $\sum_{i=1}^{3} \left(\frac{\Delta \bar{x}^i}{\Delta \bar{x}^4} \right)^2 = (\Lambda^4_4)^{-2}[(\Lambda^4_4)^2 - 1]$. Physically, we interpret these equalities as saying the velocity of $\bar{S}$ relative to S and the velocity of S relative to $\bar{S}$ have the same constant magnitude which we shall denote by β_r. Thus, $\beta_r^2 = 1 - (\Lambda^4_4)^{-2}$ so in particular, $0 \leq \beta_r^2 < 1$ and $\beta_r = 0$ if and only if Λ is a rotation (Lemma 1.3.4). Solving for Λ^4_4 (and taking the positive square root since Λ is assumed orthochronous) yields

$$\Lambda^4_4 = (1 - \beta_r^2)^{-1/2} \quad (= \overline{\Lambda}^4_4) \tag{44}$$

The quantity $(1 - \beta_r^2)^{-1/2}$ is often designated γ. Assuming that Λ is not a rotation we may write $\vec{u}$ as

$$\vec{u} = \beta_r \vec{d} = \beta_r (d^1 e_1 + d^2 e_2 + d^3 e_3), \; d^i = u^i / \beta_r \tag{45}$$

where $\vec{d}$ is called the *direction 3-vector* of $\bar{\mathcal{S}}$ relative to $\mathcal{S}$ and the d^i are interpreted as the direction cosines of the directed line segment in Σ along which the observer in $\mathcal{S}$ sees $\bar{\mathcal{S}}$ moving. Similarly.

$$\vec{\bar{u}} = \beta_r \, \vec{\bar{d}} = \beta_r(\bar{d}^1 \bar{e}_1 + \bar{d}^2 \bar{e}_2 + \bar{d}^3 \bar{e}_3), \; \bar{d}^i = \bar{u}^i / \beta_r \; . \tag{46}$$

Exercise 1.3.5. Show that the d^i are the components of the normalized projection of $\bar{e}_4$ onto the subspace $[e_1, e_2, e_3]$ of $\mathcal{M}$ spanned by e_1, e_2 and e_3, i.e., that

$$d^i = \frac{\bar{e}_4 \cdot e_i}{\left(\sum_{j=1}^{3} (\bar{e}_4 \cdot e_j)^2 \right)^{1/2}} \; , \quad i = 1,2,3 \tag{47}$$

and similarly,

$$\bar{d}^i = \frac{e_4 \cdot \bar{e}_i}{\left(\sum_{j=1}^{3} (e_4 \cdot \bar{e}_j)^2 \right)^{1/2}} \; , \quad i = 1,2,3. \tag{48}$$

Exercise 1.3.6. Show that $\bar{e}_4 = \gamma\,(\beta_r \vec{d} + e_4)$ and, similarly,

$$e_4 = \gamma(\beta_r \vec{\bar{d}} + \bar{e}_4) \; .$$

Comparing (42) and (45) and using (44) we obtain

$$\bar{\Lambda}^i_4 = -\Lambda^4_i = \beta_r(1 - \beta_r^2)^{-1/2} \, d^i \;\; , \;\; i = 1,2,3 \tag{49}$$

and similarly

$$\Lambda^i_4 = -\bar{\Lambda}^4_i = \beta_r(1 - \beta_r^2)^{-1/2} \, \bar{d}^i \;\; , \;\; i = 1,2,3. \tag{50}$$

Equations (44), (49) and (50) give the last row and column of Λ in terms of physically measurable quantities. Even at this stage a number of interesting kinematic consequenes become apparent. From (8) we obtain

$$\Delta\bar{x}^4 = -\beta_r \gamma \, (d^1 \Delta x^1 + d^2 \Delta x^2 + d^3 \Delta x^3) + \gamma \Delta x^4 \tag{51}$$

for any two events. Let us consider the special case of two events on the worldline of a *point at rest in* $\mathcal{S}$. Then $\Delta x^1 = \Delta x^2 = \Delta x^3 = 0$ so

$$\Delta \bar{x}^4 = \gamma \Delta x^4 = \frac{1}{\sqrt{1-\beta_r^2}} \Delta x^4 \ . \tag{52}$$

In particular, $\Delta \bar{x}^4 = \Delta x^4$ if and only if Λ is a rotation. Any relative motion of $\mathcal{S}$ and $\bar{\mathcal{S}}$ gives rise to a *time dilation* effect according to which $\Delta \bar{x}^4 > \Delta x^4$. Since our two events can be interpreted as two readings on one of the clocks in $\mathcal{S}$, $\bar{\mathcal{S}}$ will conclude that the clocks in $\mathcal{S}$ are running slow (even though they are, by assumption, identical).

Exercise 1.3.7. Show that the time dilation effect is entirely symmetrical, i.e., that for two events with $\Delta \bar{x}^1 = \Delta \bar{x}^2 = \Delta \bar{x}^3 = 0$,

$$\Delta x^4 = \gamma \Delta \bar{x}^4 = \frac{1}{\sqrt{1-\beta_r^2}} \Delta \bar{x}^4 \ . \tag{53}$$

This time dilation effect is in no sense an illusion; it is quite "real" and can manifest itself in observable phenomena. One such instance occurs in the study of cosmic rays ("showers" of various kinds of elementary particles from outer space which impact the earth). Certain types of mesons which are encountered in cosmic radiation are so short-lived (at rest) that, even if they could travel at the speed of light (which they cannot, of course) the time required to traverse our atmosphere would be some ten times their normal life span. They should not be able to reach the earth. But they do! Time dilation in a sense "keeps them young". The meson's notion of time is not the same as ours. What seems a normal lifetime to the meson appears much longer to us. It is well to keep in mind also that we have been rather vague about what we mean by a "clock". Essentially any phenomenon involving observable change (successive readings on a Timex, vibrations of an atom, the lifetime of a meson, or a human being) is a "clock" and is therefore subject to the effects of time dilation. Of course, these effects will be negligibly small unless β_r is quite close to 1 (the speed of light).

Another special case of (51) is of interest. Let us suppose that our two events are judged *simultaneous in* $\mathcal{S}$, i.e., that $\Delta x^4 = 0$ so

$$\Delta \bar{x}^4 = -\beta_r \gamma (d^1 \Delta x^1 + d^2 \Delta x^2 + d^3 \Delta x^3) \ . \tag{54}$$

Again assuming that $\beta_r \neq 0$ we find that, in general, $\Delta \bar{x}^4$ will *not* be zero, i.e., that the two events will not be judged simultaneous in $\bar{\mathcal{S}}$. Indeed, $\mathcal{S}$ and $\bar{\mathcal{S}}$ will agree on the simultaneity of these two events if and only if the spatial locations of the events in Σ bear a very special relation to the direction in Σ along which $\bar{\Sigma}$ is moving, namely,

$$d^1 \Delta x^1 + d^2 \Delta x^2 + d^3 \Delta x^3 = 0 \tag{55}$$

(the displacement vector in Σ between the locations of the two events is perpendicular to the direction of $\bar{\Sigma}$'s motion in Σ). Otherwise $\Delta\bar{x}^4 \neq 0$ and we have an instance of the *relativity of simultaneity* to which we referred in section 1.2.

It will be useful at this point to isolate a certain subgroup of the Lorentz group $\mathcal{L}$ which contains all of the physically significant information about Lorentz transformations, but has much of the unimportant detail pruned away. We do this in the obvious way by assuming that the spatial axes of $\mathcal{S}$ and $\bar{\mathcal{S}}$ have a particularly simple relative orientation. Specifically, we consider the special case in which the direction cosines d^i and $\bar{d}^i$, $i = 1,2,3$, are given by $d^1 = 1, \bar{d}^1 = -1$ and $d^2 = \bar{d}^2 = d^3 = \bar{d}^3 = 0$. Thus, the direction vectors are $\vec{d} = e_1$ and $\vec{\bar{d}} = -\bar{e}_1$. Physically, this corresponds to the situation in which an observer in $\mathcal{S}$ sees $\bar{\mathcal{S}}$ moving in the direction of the positive x^1-axis and an observer in $\bar{\mathcal{S}}$ sees $\mathcal{S}$ moving in the direction of the negative $\bar{x}^1$-axis. Since the origins of the spatial coordinate systems of $\mathcal{S}$ and $\bar{\mathcal{S}}$ coincided at $x^4 = \bar{x}^4 = 0$, we picture the motion of these two systems as being along their common x^1–, $\bar{x}^1$-axis. Now, from (44), (49) and (50) we find that the transformation matrix Λ must have the form

$$\Lambda = \begin{bmatrix} \Lambda^1_1 & \Lambda^1_2 & \Lambda^1_3 & -\beta_r\gamma \\ \Lambda^2_1 & \Lambda^2_2 & \Lambda^2_3 & 0 \\ \Lambda^3_1 & \Lambda^3_2 & \Lambda^3_3 & 0 \\ -\beta_r\gamma & 0 & 0 & \gamma \end{bmatrix} \tag{56}$$

Exercise 1.3.8. Use the orthogonality conditions (3) and (4) to show that Λ must take the form

$$\Lambda = \begin{bmatrix} \gamma & 0 & 0 & -\beta_r\gamma \\ 0 & \Lambda^2_2 & \Lambda^2_3 & 0 \\ 0 & \Lambda^3_2 & \Lambda^3_3 & 0 \\ -\beta_r\gamma & 0 & 0 & \gamma \end{bmatrix} \tag{57}$$

where $[\Lambda^i_j]_{i,j=2,3}$ is a 2×2 unimodular orthogonal matrix, i.e., a rotation of the plane $\mathbb{R}^2$.

To discover the differences between these various elements of $\mathcal{L}$ we consider first the simplest possible choice for the 2×2 unimodular orthogonal matrix, i.e., the identity matrix. The corresponding Lorentz transformation is

$$\Lambda = \begin{bmatrix} \gamma & 0 & 0 & -\beta_r\gamma \\ 0 & 1 & 0 & 0 \\ 0 & 0 & 1 & 0 \\ -\beta_r\gamma & 0 & 0 & \gamma \end{bmatrix} \tag{58}$$

and its associated coordinate transformation is

$$\begin{aligned} \bar{x}^1 &= (1-\beta_r^2)^{-1/2}\, x^1 - \beta_r(1-\beta_r^2)^{-1/2}\, x^4 \\ \bar{x}^2 &= x^2 \\ \bar{x}^3 &= x^3 \\ \bar{x}^4 &= -\beta_r(1-\beta_r^2)^{-1/2}\, x^1 + (1-\beta_r^2)^{-1/2}\, x^4 \end{aligned} \tag{59}$$

By virtue of the equalities $\bar{x}^2 = x^2$ and $\bar{x}^3 = x^3$ the physical relationship between the spatial axes in $\mathcal{S}$ and $\bar{\mathcal{S}}$ is as shown below:

Figure 1.1

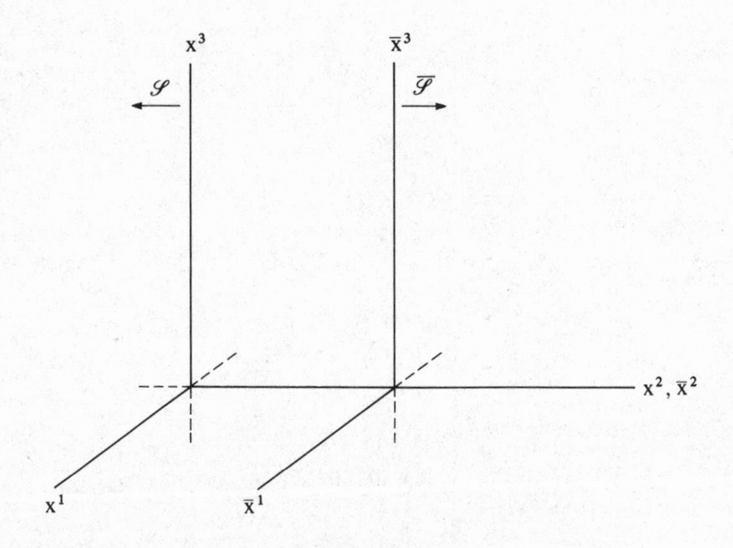

Remark. These spatial axes are said to be in *standard configuration.*

Now it is clear that any Lorentz transformation of the form (57) will correspond to the physical situation in which the $\bar{x}^2$- and $\bar{x}^3$-axes of $\bar{\mathcal{S}}$ are rotated in their own plane from the position shown in Figure 1.5.

By (30) the inverse of the Lorentz transformation Λ defined by (58) is

$$\Lambda^{-1} = \begin{bmatrix} \gamma & 0 & 0 & \beta_r\gamma \\ 0 & 1 & 0 & 0 \\ 0 & 0 & 1 & 0 \\ \beta_r\gamma & 0 & 0 & \gamma \end{bmatrix} \tag{60}$$

and the corresponding inverse coordinate transformation is

$$\begin{aligned} x^1 &= (1-\beta_r^2)^{-1/2}\,\bar{x}^1 + \beta_r(1-\beta_r^2)^{-1/2}\,\bar{x}^4 \\ x^2 &= \bar{x}^2 \\ x^3 &= \bar{x}^3 \\ x^4 &= \beta_r(1-\beta_r^2)^{-1/2}\,\bar{x}^1 + (1-\beta_r^2)^{-1/2}\,\bar{x}^4 \end{aligned} \tag{61}$$

Any Lorentz transformation Λ of the form (58) or (60), i.e., with $\Lambda^2_4 = \Lambda^3_4 = \Lambda^4_2 = \Lambda^4_3 = 0$ and $[\Lambda^i_j]_{i,j=2,3} = id_{2\times 2}$ is called a *special Lorentz transformation.* Since Λ and Λ^{-1} differ only in the signs of the (1,4) and (4,1) entries it is customary, when discussing special Lorentz transformations, to allow $-1 < \beta_r < 1$. By choosing $\beta_r > 0$ when $\Lambda^1_4 < 0$ and $\beta_r < 0$ when $\Lambda^1_4 > 0$ all special Lorentz transformations can be written in the form (58) and we shall henceforth adopt this convention. For each real number β with $-1 < \beta < 1$ we define $\gamma = \gamma(\beta) = (1-\beta^2)^{-1/2}$ and

$$\Lambda(\beta) = \begin{bmatrix} \gamma & 0 & 0 & -\beta\gamma \\ 0 & 1 & 0 & 0 \\ 0 & 0 & 1 & 0 \\ -\beta\gamma & 0 & 0 & \gamma \end{bmatrix}$$

Exercise 1.3.9. Show that if $-1 < \beta_1 \le \beta_2 < 1$, then

$$\Lambda(\beta_2)\Lambda(\beta_1) = \Lambda\left(\frac{\beta_1 + \beta_2}{1 + \beta_1\beta_2}\right) \tag{62}$$

By referring the three special Lorentz transformations $\Lambda(\beta_1)$, $\Lambda(\beta_2)$ and $\Lambda(\beta_2)\,\Lambda(\beta_1)$ to the corresponding admissible frames of reference one arrives at the following physical interpretation of (62): If the speed of $\bar{\mathcal{S}}$ relative to $\mathcal{S}$ is β_1 and the speed of $\bar{\bar{\mathcal{S}}}$ relative to $\bar{\mathcal{S}}$ is β_2, then the speed of $\bar{\bar{\mathcal{S}}}$ relative to $\mathcal{S}$ is not $\beta_1 + \beta_2$ as one might expect, but rather

$$\frac{\beta_1 + \beta_2}{1 + \beta_1\beta_2} \tag{63}$$

which is always *less* than $\beta_1 + \beta_2$ provided $\beta_1\beta_2 \neq 0$. (62) is generally known as the *relativistic addition of velocities formula.*

Exercise 1.3.10. Show that if $|\beta_1| < 1$ and $|\beta_2| < 1$, then

$$\left| \frac{\beta_1 + \beta_2}{1 + \beta_1\beta_2} \right| < 1 \ .$$

The non-additivity of velocities in relativity is to be expected, of course, since additivity would imply that arbitrarily large speeds could be attained relative to an admissible frame. Nevertheless, it is often convenient to measure speeds with an alternative "velocity parameter" θ that *is* additive. An analogous situation occurs in plane Euclidean geometry where one has the option of describing the relative orientation of two Cartesian coordinate systems by means of angles (which are additive) or slopes (which are not). What we would like then is a measure θ of relative velocities with the property that if θ_1 is the velocity parameter of $\bar{\mathcal{S}}$ relative to $\mathcal{S}$ and θ_2 is the velocity parameter of $\bar{\bar{\mathcal{S}}}$ relative to $\bar{\mathcal{S}}$, then the velocity parameter of $\bar{\bar{\mathcal{S}}}$ relative to $\mathcal{S}$ is $\theta_1 + \theta_2$. Since θ measures relative velocity, β will be some one-to-one function $f(\theta)$ of θ. Additivity and (62) require that f satisfy the functional equation

$$f(\theta_1 + \theta_2) = \frac{f(\theta_1) + f(\theta_2)}{1 + f(\theta_1)f(\theta_2)} \ . \tag{64}$$

Being suggestive of the sum formula for the hyperbolic tangent, (64) suggests the change of variable

$$\beta = \tanh\theta \quad \text{or} \quad \theta = \tanh^{-1}\beta \ . \tag{65}$$

Observe that $\tanh^{-1}$ is a one-to-one differentiable function of $(-1,1)$ onto $\mathbb{R}$ with the property that $\beta \to \pm 1$ implies $\theta \to \pm\infty$, i.e., the speed of light has infinite velocity parameter.

Exercise 1.3.11. Show that there is a *unique* differentiable function $\beta = f(\theta)$ on $\mathbb{R}$ (namely, $\tanh\theta$) which satisfies (64) and

$$\lim_{\theta \to 0} \frac{f(\theta)}{\theta} = 1$$

(for small speeds, β and θ are nearly equal). *Hint.* Show that such an f necessarily satisfies the initial value problem $f'(\theta) = 1 - (f(\theta))^2$, $f(0) = 0$ and appeal to the Uniqueness Theorem for solutions to such problems (see [Har]).

Exercise 1.3.12. Show that if $\beta = \tanh\theta$, then the *hyperbolic form* of the Lorentz transformation $\Lambda(\beta)$ is

$$L(\theta) = \begin{bmatrix} \cosh\theta & 0 & 0 & -\sinh\theta \\ 0 & 1 & 0 & 0 \\ 0 & 0 & 1 & 0 \\ -\sinh\theta & 0 & 0 & \cosh\theta \end{bmatrix}$$

Earlier we suggested that all of the physically interesting behavior of proper, orthochronous Lorentz transformations is exhibited by the special Lorentz transformations. What we had in mind was the following theorem which asserts that any element of $\mathcal{L}$ differs from some $L(\theta)$ only by at most two (physically uninteresting) rotations.

Theorem 1.3.5. Let Λ be a proper, orthochronous Lorentz transformation. Then there exists a real number θ and two rotations R_1 and R_2 in $\mathcal{R}$ such that $\Lambda = R_1 L(\theta) R_2$.

Since we shall make no use of this result we simply refer the reader to the proof in [Nai].

From the point-of-view of physics it is therefore sufficient to limit one's attention to frames whose spatial coordinate axes are in standard configuration. We shall now discuss the phenomenon of *length contraction* within this context. Thus, we consider two frames $\mathcal{S}$ and $\bar{\mathcal{S}}$ whose spatial axes are as shown in Figure 1.5. Consider also a "rigid" rod at rest relative to $\bar{\mathcal{S}}$ and lying along the $\bar{x}^1$-axis between $\bar{x}_0^1$ and $\bar{x}_1^1$. Thus, $\Delta\bar{x}^1 = \bar{x}_1^1 - \bar{x}_0^1$ is the measured length of the rod in $\bar{\mathcal{S}}$. The worldline of the rod's left-(resp., right-) hand endpoint has $\bar{\mathcal{S}}$-coordinates $(\bar{x}_0^1, 0, 0, \bar{x}^4)$ (resp., $(\bar{x}_1^1, 0, 0, \bar{x}^4)$), $-\infty < \bar{x}^4 < \infty$. $\mathcal{S}$ will measure the length of this same rod by locating its two endpoints "simultaneously", i.e., by finding one event on each of these worldlines *with the same* x^4 (*not* $\bar{x}^4$). But for any *fixed* x^4, the transformation equations (59) give $\bar{x}_0^1 = (1-\beta_r^2)^{-1/2}\, x_0^1 - \beta_r x^4$ and $\bar{x}_1^1 = (1-\beta_r^2)^{-1/2}\, x_1^1 - \beta_r x^4$ so that $\Delta\bar{x}^1 = (1-\beta_r^2)^{-1/2}\,\Delta x^1$ and therefore

$$\Delta x^1 = (1-\beta_r^2)^{1/2}\,\Delta\bar{x}^1 \ . \tag{66}$$

Since $(1-\beta_r^2)^{1/2} < 1$ we find that the measured length of the rod in $\mathcal{S}$ is *less* than its measured length in $\bar{\mathcal{S}}$ by a factor of $(1-\beta_r^2)^{1/2}$. By reversing the roles of $\mathcal{S}$ and $\bar{\mathcal{S}}$ we again find that this effect is entirely symmetrical.

Time dilation and length contraction are mathematically trivial, of course, but their physical interpretation is very often subtle and quite delicate. We recommend to the reader's attention the following well-known example of the sort of "paradox" to which one can be led by a superficial grasp of the phenomena.

Imagine a barn which, at rest, measures 8 meters in length. A (very fast) runner carries a pole of rest length 16 meters toward the barn at such a high speed that, for an observer at rest in the barn, it appears Lorentz contracted to 8 meters and therefore fits inside the barn. This observer slams the front door shut at the instant the back of the pole enters the front of the barn and so contains the pole entirely within the barn. But is it not true that, by symmetry, the runner sees the barn Lorentz contracted to 4 meters so that the 16 meter pole could never fit entirely within it?

Exercise 1.3.13. Resolve the difficulty. *Hint.* Let $\mathcal{S}$ and $\bar{\mathcal{S}}$ respectively denote the rest frames of the barn and the pole and assume these frames are related by (59) and (61). Calculate β_r. Suppose that the front of the pole enters the front of the barn at (0,0,0,0). Now consider the two events at which the front of the pole hits the back of the barn and the back of the pole enters the front of the barn. Finally, think about the maximum speed at which the signal to stop can be communicated from the front to the back of the pole.

1.4. Null Vectors and Photons

Consider two distinct events x_0 and x for which the *displacement vector* $x - x_0$ from x_0 to x is null, i.e., $Q(x - x_0) = 0$. Relative to any admissible basis,

$$(x^1 - x_0^1)^2 + (x^2 - x_0^2)^2 + (x^3 - x_0^3)^2 - (x^4 - x_0^4)^2 = 0 \ . \tag{67}$$

Clearly, $\Delta x^4 = x^4 - x_0^4 \neq 0$ and, by Theorem 1.3.3, the sign of Δx^4 is the same in all admissible coordinate systems. We shall say that $x - x_0$ is *future-directed* if $\Delta x^4 > 0$ and *past-directed* if $\Delta x^4 < 0$. We now define a binary relation $<$ on $\mathcal{M}$ as follows: $x_0 < x$ if and only if $x - x_0$ is null and future-directed. We have already seen that, since (67) implies that the spatial separation of x_0 and x is equal to the distance light would travel during the time lapse between x_0 and x, the physical interpretation of $x_0 < x$ is that a photon "can get from x_0 to x", i.e., that x_0 and x respectively can be regarded as the emission and subsequent reception of some light signal. For any two distinct events x_0 and x with $Q(x - x_0) = 0$ we define the *light ray* $R_{x_0,x}$, through x_0 and x by

$$R_{x_0,x} = \{x_0 + t(x - x_0) : t \in \mathbb{R}\} \ .$$

Since, relative to Cartesian coordinates in $\mathbb{R}^4$, (67) is the equation of a right-circular cone with vertex x_0, we shall refer to the set

$$C_N(x_0) = \{x \in \mathcal{M} : Q(x - x_0) = 0\}$$

as the *null cone* (or *light cone*) at x_0 and picture it by suppressing the third spatial dimension x^3:

Figure 1.2

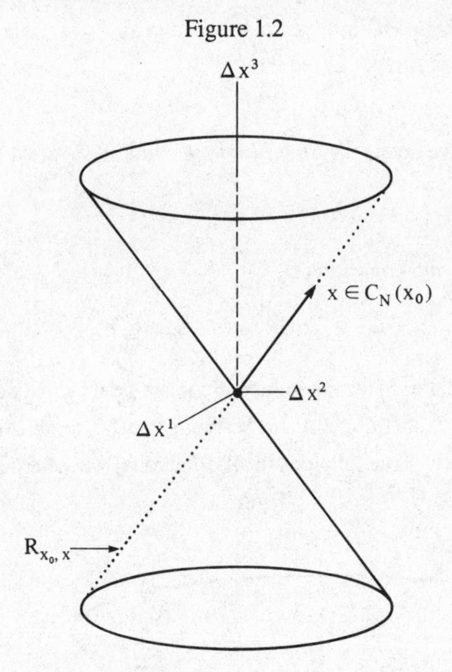

Lemma 1.4.1. Two nonzero null vectors x and y in $\mathcal{M}$ are orthogonal if and only if they are parallel, i.e., iff there is a t in $\mathbb{R}$ such that $x = ty$.

Exercise 1.4.1. Prove Lemma 1.4.1. *Hint.* For the necessity recall when equality holds in the Schwartz Inequality for $\mathbb{R}^3$.

$C_N(x_0)$ is just the union of all the light rays through x_0. Indeed,

Proposition 1.4.2. Let x_0 and x be two distinct events with $Q(x - x_0) = 0$. Then

$$R_{x_0,x} = C_N(x) \cap C_N(x_0) \ . \tag{68}$$

Proof. First let $z = x_0 + t(x - x_0)$ be an element of $R_{x_0,x}$. Then $z - x_0 = t(x - x_0)$ so $Q(z - x_0) = t^2 Q(x - x_0) = 0$ so z is in $C_N(x_0)$. Similarly, $z \in C_N(x)$. To prove the reverse containment we assume z is in $C_N(x) \cap C_N(x_0)$. Then each of the vectors $z - x$, $z - x_0$ and $x_0 - x$ is null. But $z - x_0 = (z - x) - (x_0 - x)$ so

$0 = Q(z - x_0) = (z - x)^2 - 2(z - x) \cdot (x_0 - x) + (x_0 - x)^2 = -2(z - x) \cdot (x_0 - x)$. Thus, $(z - x) \cdot (x_0 - x) = 0$. If $z = x$ we are done. If $z \neq x$, then, since $x_0 \neq x$, we may apply Lemma 1.4.1 to the orthogonal null vectors $z - x$ and $x_0 - x$ to obtain a t in $\mathbb{R}$ such that $z - x = t(x_0 - x)$. Thus $z \in R_{x_0,x}$ as required. Q.E.D.

For any x_0 in $\mathcal{M}$ we define the *future* (or *upper*) *null cone* at x_0 by

$$C_N^+(x_0) = \{x \in \mathcal{M} : x_0 < x\}$$

and the *past* (or *lower*) *null cone* at x_0 by

$$C_N^-(x_0) = \{x \in \mathcal{M} : x < x_0\} .$$

Thus, x is in $C_N^+(x_0)$ iff a light signal emitted at x_0 can be received at x. x is in $C_N^-(x_0)$ iff x_0 is in $C_N^+(x)$. Observe that $C_N^+(x_0)$ can be thought of as the history in spacetime of a spherical electromagnetic wave (photons in all directions) whose emission event is x_0:

Figure 1.3

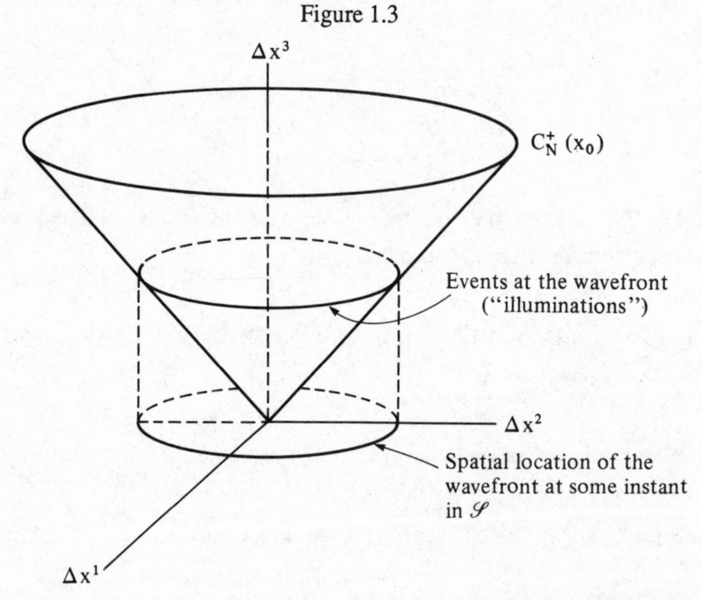

Exercise 1.4.2. Construct an example to show that the relation $<$ is *not* transitive (i.e., that $x < y$ and $y < z$ does not imply $x < z$) and interpret this non-transitivity physically.

1.5. Timelike Vectors and Material Particles

Next we consider two events x and x_0 for which $Q(x - x_0) < 0$, i.e., $x - x_0$ is timelike. Relative to any admissible basis $\{e_a\}$ we have $(\Delta x^1)^2 + (\Delta x^2)^2 + (\Delta x^3)^2 < (\Delta x^4)^2$. Obviously, $\Delta x^4 \neq 0$ and we can assume without loss of generality that $\Delta x^4 > 0$. Thus, we obtain

$$\frac{((\Delta x^1)^2 + (\Delta x^2)^2 + (\Delta x^3)^2)^{1/2}}{\Delta x^4} < 1 \ . \tag{69}$$

Physically it is therefore clear (remember that the speed of light is 1!) that if one were to move with speed $\frac{((\Delta x^1)^2 + (\Delta x^2)^2 + (\Delta x^3)^2)^{1/2}}{\Delta x^4}$ relative to the frame S corresponding to $\{e_a\}$ along the line in Σ from (x_0^1, x_0^2, x_0^3) to (x^1, x^2, x^3) and if one were present at x_0, then one would also experience x, i.e., there is an admissible frame of reference $\bar{S}$ in which x_0 and x occur at the same spatial point, one after the other. Specifically, we will show that if we choose $\beta_r = ((\Delta x^1)^2 + (\Delta x^2)^2 + (\Delta x^3)^2)^{1/2} / \Delta x^4$ and let d^1, d^2 and d^3 be the direction cosines in Σ of the directed line segment from (x_0^1, x_0^2, x_0^3) to (x^1, x^2, x^3), then the basis $\{\bar{e}_a\}$ for $\mathcal{M}$ obtained from $\{e_a\}$ by performing a Lorentz transformation whose fourth row is $\Lambda^4_i = -\beta_r(1 - \beta_r^2)^{-1/2}\, d^i$, $i = 1,2,3$, $\Lambda^4_4 = (1 - \beta_r^2)^{-1/2} = \gamma$ (see the Remark below) has the property that $\Delta\bar{x}^1 = \Delta\bar{x}^2 = \Delta\bar{x}^3 = 0$. To prove this we compute $\Delta\bar{x}^4 = \Lambda^4_b\,\Delta x^b$. To simplify the computations, let $\Delta\vec{x} = ((\Delta x^1)^2 + (\Delta x^2)^2 + (\Delta x^3)^2)^{1/2}$. Then $\beta_r^2 = \Delta\vec{x}^2 / (\Delta x^4)^2$, $\gamma = \Delta x^4 / \sqrt{-Q(x - x_0)}$, $\beta_r\,\gamma = \Delta\vec{x} / \sqrt{-Q(x - x_0)}$ and $d^i = \Delta x^i / \Delta\vec{x}$ for $i = 1,2,3$ (we may clearly assume that $\Delta\vec{x} \neq 0$ for otherwise there is nothing to prove). Thus,

$$\begin{aligned}
\Delta\bar{x}^4 &= \Lambda^4_1\Delta x^1 + \Lambda^4_2\Delta x^2 + \Lambda^4_3\Delta x^3 + \Lambda^4_4\,\Delta x^4 \\
&= -\beta_r\gamma\,(d^1\Delta x^1 + d^2\Delta x^2 + d^3\Delta x^3) + \gamma\,\Delta x^4 \\
&= -\frac{\Delta\vec{x}}{\sqrt{-Q(x - x_0)}}\,(\Delta\vec{x}) + \frac{(\Delta x^4)^2}{\sqrt{-Q(x - x_0)}} = \sqrt{-Q(x - x_0)}\ .
\end{aligned}$$

Consequently, $Q(x - x_0) = -(\Delta\bar{x}^4)^2$. But, by computing $Q(x - x_0)$ relative to the basis $\{\bar{e}_a\}$ we find that $Q(x - x_0) = (\Delta\bar{x}^1)^2 + (\Delta\bar{x}^2)^2 + (\Delta\bar{x}^3)^2 - (\Delta\bar{x}^4)^2$ so we must have $(\Delta\bar{x}^1)^2 + (\Delta\bar{x}^2)^2 + (\Delta\bar{x}^3)^2 = 0$, i.e., $\Delta\bar{x}^1 = \Delta\bar{x}^2 = \Delta\bar{x}^3 = 0$.

For any two events x and x_0 for which $x - x_0$ is timelike we define the *duration* $\tau(x - x_0)$ of $x - x_0$ by

$$\tau(x - x_0) = \sqrt{-Q(x - x_0)}$$

If the vector $x = x - 0$ is timelike we generally write $\tau(x-0) = \tau(x) = \sqrt{-Q(x)}$. We have just seen that $\tau(x - x_0)$ *is to be interpreted physically as the time separation of* $\mathbf{x}$ *and* $\mathbf{x_0}$ *in any admissible frame of reference in which both events occur at the same spatial point.*

Remark. Given β_r, d^1, d^2 and d^3 there will, in general, be many Lorentz transformations whose fourth row is $\Lambda^4_i = -\beta_r(1-\beta_r^2)^{-1/2}\, d^i$, $i = 1,2,3$, $\Lambda^4_4 = (1-\beta_r^2)^{-1/2} = \gamma$ as required in the previous argument. One specific choice for the remaining entries which is often found to be convenient is as follows: $\Lambda^i_4 = -\beta_r \gamma d^i$, $i = 1,2,3$, $\Lambda^i_j = (\gamma - 1)d^i d^j + \delta^i_j$, $i,j = 1,2,3$, where δ^i_j is the Kronecker delta.

Exercise 1.5.1. Show that the matrix $[\Lambda^a_b]_{a,b=1,2,3,4}$ thus defined is indeed an element of $\mathcal{L}$ and describe the physical relationship between the spatial coordinate axes of two admissible frames of reference related by it.

A subset of $\mathcal{M}$ of the form $\{x_0 + t(x - x_0) : t \in \mathbb{R}\}$, where $Q(x - x_0) < 0$, is called a *timelike straight line.*

Exercise 1.5.2. Show that for any timelike straight line, there is an admissible frame of reference in which all the events on the line occur at the same spatial point.

A timelike straight line which passes through the origin is called a *time axis.* We claim that *any time axis* $\boldsymbol{T}$ *can be identified with the set of events in the history of some admissible observer,* i.e., with the x^4 axis of some admissible coordinatization of $\mathcal{M}$. To see this we select an event $\hat{e}_4$ on T with $\hat{e}_4 \cdot \hat{e}_4 = -1$ and let $[\hat{e}_4]$ be the linear span of $\hat{e}_4$ in $\mathcal{M}$. As point-sets, $[\hat{e}_4] = T$. Next let $[\hat{e}_4]^\perp = \{x \in \mathcal{M} : x \cdot \hat{e}_4 = 0\}$ be the *orthogonal complement* of $[\hat{e}_4]$ in $\mathcal{M}$. $[\hat{e}_4]^\perp$ is clearly also a linear subspace of $\mathcal{M}$. We claim that $\mathcal{M} = [\hat{e}_4] \oplus [\hat{e}_4]^\perp$. Indeed, let x be an element of $\mathcal{M}$ and consider the vector $v = x + (x \cdot \hat{e}_4)\,\hat{e}_4$ in $\mathcal{M}$. Since $v \cdot \hat{e}_4 = x \cdot \hat{e}_4 + (x \cdot \hat{e}_4)\,(\hat{e}_4 \cdot \hat{e}_4) = 0$, we find that v is in $[\hat{e}_4]^\perp$. Thus, the expression $x = -(x \cdot \hat{e}_4)\hat{e}_4 + v$ implies that $\mathcal{M} = [\hat{e}_4] + [\hat{e}_4]^\perp$. The sum is direct since every nonzero vector in $[\hat{e}_4]$ is timelike, while, by Corollary 1.3.2, every nonzero vector in $[\hat{e}_4]^\perp$ is spacelike. Thus, $\mathcal{M} = [\hat{e}_4] \oplus [\hat{e}_4]^\perp$ as required. Now, the restriction of the $\mathcal{M}$-inner product to $[\hat{e}_4]^\perp$ is positive definite so, by Theorem 1.1.1, we may select three vectors $\hat{e}_1, \hat{e}_2$ and $\hat{e}_3$ in $[\hat{e}_4]^\perp$ such that $\hat{e}_i \cdot \hat{e}_j = \delta_{ij}$, $i,j = 1,2,3$. Thus, $\{\hat{e}_1, \hat{e}_2, \hat{e}_3, \hat{e}_4\}$ is an orthonormal basis for $\mathcal{M}$.

Exercise 1.5.3. Fix an admissible basis $\{e_a\}$ for $\mathcal{M}$. Show that if the Lorentz transformation which carries $\{e_a\}$ onto $\{\hat{e}_a\}$ is improper, or nonorthochronous, or both, then one can multiply selected $\hat{e}_a$'s by -1 to obtain an admissible basis $\{\bar{e}_a\}$ with $[\bar{e}_4] = T$.

Observe also that if x_0 and x are two events and T is a time axis, then $x - x_0$ *is orthogonal to* **T** *iff* **x** *and* $\mathbf{x_0}$ *are simultaneous in any reference frame whose* $\mathbf{x}^4$*-axix is* T (if $T = [\bar{e}_4]$, then $(x - x_0) \cdot \bar{e}_4 = -\bar{x}^4 + \bar{x}_0^4 = \Delta \bar{x}^4$). An arbitrary timelike straight line is identified with the set of events in the history an "assistant" to some admissible observer.

Exercise 1.5.4. Show that if $x - x_0$ is timelike and s is an arbitrary non-negative real number, then there is an admissible frame in which the spatial separation of x_0 and x is s. Show also that the time separation of x_0 and x can assume any real value greater than or equal to $\tau(x - x_0)$. *Hint.* Begin with a basis $\{e_a\}$ in which $\Delta x^1 = \Delta x^2 = \Delta x^3 = 0$ and $\Delta x^4 = \tau(x - x_0)$. Now perform the special Lorentz transformation (59), where $-1 < \beta_r < 1$ is arbitrary.

Since $\tau(x - x_0)$ is a lower bound for the temporal separation of x_0 and x it is often called their *proper time separation*; when no reference to the specific events under consideration is required $\tau(x - x_0)$ will generally be denoted $\Delta \tau$.

If $x - x_0$ is timelike, then $(\Delta x^1)^2 + (\Delta x^2)^2 + (\Delta x^3)^2 < (\Delta x^4)^2$ in each admissible coordinatization, i.e., $x - x_0$ is *inside* the null cone at x_0. We define the *time cone at* x_0 by $C_T(x_0) = \{x \in \mathcal{M}\colon Q(x - x_0) < 0\}$. Clearly $\Delta x^4 \neq 0$ for such a vector and, by Theorem 1.3.3, all admissible observers agree on the sign of Δx^4. We shall say the $x - x_0$ is *future-directed* if $\Delta x^4 > 0$ and *past-directed* if $\Delta x^4 < 0$. We define another binary relation $\ll$ on $\mathcal{M}$ as follows: $x_0 \ll x$ if and only if the displacement vector $x - x_0$ from x_0 to x is timelike and future-directed. Now the *future* (or *upper*) *time cone at* x_0 is $C_T^+(x_0) = \{x \in \mathcal{M}\colon x_0 \ll x\}$. The *past* (or *lower*) *time cone at* x_0 is $C_T^-(x) = \{x \in \mathcal{M}\colon x \ll x_0\}$.

Lemma 1.5.1. The sum of any finite number of vectors in $\mathcal{M}$ all of which are timelike or null and all future-directed, is timelike and future-directed except when all of the vectors are null and parallel, in which case the sum is null and future-directed.

Proof. Sums of future-directed vectors are obviously future-directed. The remaining assertions are proved by combining three special cases.

First suppose $x_1,...,x_n$ are all timelike and future-directed. We claim that $x_1 + ... + x_n$ is timelike. By induction, it suffices to consider the case $n = 2$. Then $(x_1 + x_2) \cdot (x_1 + x_2) = x_1 \cdot x_1 + 2x_1 \cdot x_2 + x_2 \cdot x_2$ which is less than zero since x_1 and x_2 are timelike and have the same time orientation (Theorem 1.3.1).

Exercise 1.5.5. Show that if x_1 is timelike and x_2 is null and both are future-directed, then $x_1 + x_2$ is timelike.

Finally, suppose $x_1,...,x_n$ are null and future-directed. We show that $x_1 + ... + x_n$ is timelike except when all of the x_i are parallel (in which case the sum is obviously null). Induction and Exercise 1.5.5 reduce the problem to proving this when $n = 2$. Then $(x_1 + x_2) \cdot (x_1 + x_2) = 2x_1 \cdot x_2$. By Lemma 1.4.1, $x_1 \cdot x_2 = 0$ iff x_1 and x_2 are parallel. Suppose then that x_1 and x_2 are not parallel. For each $n = 1,2,3,...$ we define a y_n in $\mathcal{M}$ as follows: Choose an admissible basis $\{e_a\}$ for $\mathcal{M}$ and let $x_1 = x_1^a e_a$ and $x_2 = x_2^a e_a$. Set $y_n = x_1^1 e_1 + x_1^2 e_2 + x_1^3 e_3 + (x_1^4 + \frac{1}{n})e_4$. Each y_n is timelike and future-directed. By Thorem 1.3.1, $0 > y_n \cdot x_2 = x_1 \cdot x_2 - \frac{1}{n} x_2^4$ so $x_1 \cdot x_2 < \frac{1}{n} x_2^4$ for each n. Thus, $x_1 \cdot x_2 \leq 0$. But $x_1 \cdot x_2 \neq 0$ by assumption so $x_1 \cdot x_2 < 0$. Thus, $(x_1 + x_2) \cdot (x_1 + x_2) < 0$ as required. Q.E.D.

By considering $-x_1,...,-x_n$ rather than $x_1,...,x_n$ it follows that Lemma 1.5.1 remains true if "future-directed" is replaced everywhere by "past-directed".

Exercise 1.5.6. Show that, unlike $<$, the relation $\ll$ is transitive.

Another simple consequence of Lemma 1.5.1 is the analogue of Theorem 1.3.1 for nonzero, nonparallel null vectors.

Theorem 1.5.2. Let x and y be nonzero, nonparallel null vectors. Then x and y have the same time orientation (i.e. are both future-directed or both past-directed) iff $x \cdot y < 0$.

Proof. Suppose first that x and y have the same time orientation. Then by Lemma 1.5.1 (and the remark immediately following its proof), $x + y$ is timelike so $0 > (x + y) \cdot (x + y) = 2x \cdot y$ and therefore $x \cdot y < 0$. Conversely, if x and y have opposite time orientations, then x and $-y$ have the same time orientation so $x \cdot (-y) < 0$ and therefore $x \cdot y > 0$. Q.E.D.

If two events are such that either $x < y$ or $x \ll y$, then (physically) x is capable of influencing y either through the propagation of some electromagnetic effect (if $x < y$) or by virtue of some material phenomenon which is initiated at x (if $x \ll y$). For this reason these two relations are called *causality relations*. It is interesting to observe that each of these relations can be defined in terms of the other:

Exercise 1.5.7 Show that

(a) $x < y$ iff $x \ll y$ and $y \ll z$ imply $x \ll z$, and

(b) $x \ll y$ iff $x < y$ and $x < z < y$ for some z in $\mathcal{M}$.

A one-to-one mapping F of $\mathcal{M}$ onto itself is called a *causal automorphism* if both F and F^{-1} preserve the relation $<$, i.e., if $x < y$ iff $F(x) < F(y)$ for all x and y in $\mathcal{M}$. It follows from Exercise 1.5.7 that F is causal automorphism iff it preserves the relation $\ll$. A few examples of causal automorphisms are obvious: *Translations*, i.e., maps of the form $x \to x + x_0$ for some fixed x_0 in $\mathcal{M}$. *Dilations*, i.e., maps $x \to kx$ for some positive constant k. *Orthochronous orthogonal transformations,* i.e., orthogonal transformations $L: \mathcal{M} \to \mathcal{M}$ which satisfy $x \cdot Lx < 0$ for all timelike x in $\mathcal{M}$. Any composition of such maps is another example. Since there is nothing in the definition of a causal automorphism to suggest that such a map is necessarily affine (or even continuous!) one might expect to find a great many more examples. Nevertheless, Zeeman [Z1] has shown that we have just enumerated them all.

Theorem 1.5.3. The set of causal automorphisms of $\mathcal{M}$ coincides with the set of all compositions of translations, dilations and orthochronous orthogonal transformations.

The deeper geometrical properties of $\mathcal{M}$ (and the corresponding physical effects) depend in a crucial way upon the following two "indefinite" analogues of the Schwartz and Triangle Inequalities.

Theorem 1.5.4 (Reversed Schwartz Inequality). If x and y are timelike vectors in $\mathcal{M}$, then

$$(x \cdot y)^2 \geq x^2 y^2 \tag{70}$$

and equality holds only when x and y are linearly dependent.

Proof. Consider the vector $u = ax - by$, where $a = x \cdot y$ and $b = x \cdot x = x^2$. Observe that $u \cdot x = ax^2 - bx \cdot y = x^2(x \cdot y) - x^2(x \cdot y) = 0$. Since x is timelike, Corollary 1.3.2 implies that u is either zero or spacelike. Thus, $0 \le u^2 = a^2x^2 + b^2y^2 - 2abx \cdot y$, with equality holding only if $u = 0$. Consequently, $2abx \cdot y \le a^2x^2 + b^2y^2$, i.e.,

$$2x^2(x \cdot y)^2 \le x^2(x \cdot y)^2 + (x^2)^2y^2$$

$$2(x \cdot y)^2 \ge (x \cdot y)^2 + x^2y^2 \qquad (\text{since } x^2 < 0)$$

$$(x \cdot y)^2 \ge x^2y^2$$

and equality holds only if $u = 0$. But $u = 0$ implies $ax - by = 0$ which, since $a = x \cdot y \neq 0$ by Theorem 1.3.1 or $b = x \cdot x < 0$, implies that x and y are linearly dependent. Conversely, if x and y are linearly dependent, then one is a multiple of the other and equality obviously holds in (70). Q.E.D.

Theorem 1.5.5 (Reversed Triangle Inequality). Let x and y be timelike vectors with the same time orientation (i.e., $x \cdot y < 0$).
Then

$$\tau(x + y) \ge \tau(x) + \tau(y) \tag{71}$$

and equality holds only when x and y are linearly dependent.

Proof. By Theorem 1.5.4, $(x \cdot y)^2 \ge x^2y^2 = (-x^2)(-y^2)$ so $|x \cdot y| \ge \sqrt{-x^2}\,\sqrt{-y^2}$. But $x \cdot y < 0$ so we must have $x \cdot y \le -\sqrt{-x^2}\,\sqrt{-y^2}$ and therefore

$$-2x \cdot y \ge 2\sqrt{-x^2}\,\sqrt{-y^2}\ . \tag{72}$$

Now, $-(x + y)^2 = -(x + y) \cdot (x + y) = -x^2 - 2x \cdot y - y^2 \ge -x^2 + 2\sqrt{-x^2}\,\sqrt{-y^2} - y^2$ by (72). Thus,

$$-(x + y)^2 \ge (\sqrt{-x^2} + \sqrt{-y^2})^2\ .$$

$$\sqrt{-(x + y)^2} \ge \sqrt{-x^2} + \sqrt{-y^2}$$

$$\sqrt{-Q(x + y)} \ge \sqrt{-Q(x)} + \sqrt{-Q(y)}$$

$$\tau(x + y) \ge \tau(x) + \tau(y)$$

as required. If equality holds in (71), then, by reversing the preceding steps, we obtain $-2x \cdot y = 2\sqrt{-x^2}\,\sqrt{-y^2}$ and therefore $(x \cdot y)^2 = x^2 y^2$ so, by Theorem 1.5.4., x and y are linearly dependent. Q.E.D.

The reason that the sense of the inequality in Theorem 1.5.5 is "reversed" becomes particularly transparent by choosing a coordinate system relative to which $x = (x^1, x^2, x^3, x^4)$, $y = (y^1, y^2, y^3, y^4)$ and $x + y = (0,0,0,x^4 + y^4)$ (this simply amounts to taking the time axis through $x + y$ as the x^4-axis. For then $\tau(x) = ((x^4)^2 - (x^1)^2 - (x^2)^2 - (x^3)^2)^{1/2} < x^4$ and $\tau(y) < y^4$, but $\tau(x + y) = x^4 + y^4$. Observe also that, by Lemma 1.5.1, the Reversed Triangle Inequality extends to arbitrary finite sums of timelike vectors with the same time orientation.

In order to discuss worldlines of material particles we shall require a few preliminary definitions. Let $I \subseteq \mathbb{R}$ be an interval. A map $\alpha : I \to \mathcal{M}$ is called a *curve* in $\mathcal{M}$. Relative to any admissible coordinate system for $\mathcal{M}$ we can write $\alpha(t) = (x^1(t), x^2(t), x^3(t), x^4(t))$ for each t in I. We shall assume that α is *smooth*, i.e., that each $x^a(t)$ is infinitely differentiable (C^∞) and that the *velocity vector* $\alpha'(t) = \left(\frac{dx^1}{dt}, \frac{dx^2}{dt}, \frac{dx^3}{dt}, \frac{dx^4}{dt}\right)$ is nonzero for each t in I. Observe that if $\bar{x}^a = \Lambda^a{}_b x^b$, then $\frac{d\bar{x}^a}{dt} = \Lambda^a{}_b \frac{dx^b}{dt}$ so this definition of smoothness does not depend on our choice of admissible coordinates (recall the $[\Lambda^a{}_b]$ is nonsingular). α is said to be *timelike* if $\alpha'(t)$ is timelike for each t in I, i.e., if $\alpha'(t) \cdot \alpha'(t) = \eta_{ab} \frac{dx^a}{dt} \frac{dx^b}{dt} < 0$, and *future-directed* if $\frac{dx^4}{dt} > 0$ for every t. A smooth, future-directed, timelike curve in $\mathcal{M}$ will be called a *worldline of a material particle*. If α has the form $\alpha(t) = x_0 + t(x - x_0)$, where $x_0 \ll x$, then α is the *worldline of a free material particle* (if, in this definition, we replace $x_0 \ll x$ by $x_0 < x$ we obtain a *worldline of a photon*). If $J \subseteq \mathbb{R}$ is another interval and $h : J \to I$, $t = h(s)$, is a C^∞ function with $h'(s) > 0$ for each s in J, then $\beta = \alpha \circ h : J \to \mathcal{M}$ is called a *reparametrization* of α.

Exercise 1.5.8. Show that $\beta'(s) = h'(s)\,\alpha'(h(s))$ and conclude that the definitions of "smooth", "timelike" and "future-directed" curves are all independent of parametrization.

If α: $[a,b] \to \mathcal{M}$ is the worldline of some material particle in $\mathcal{M}$,. then we define the *proper time length* of α by

$$L(\alpha) = \int_a^b |\alpha'(t) \cdot \alpha'(t)|^{1/2}\, dt = \int_a^b \sqrt{-\eta_{ab} \frac{dx^a}{dt} \frac{dx^b}{dt}}\, dt$$

Exercise 1.5.9. Show that the definition of $L(\alpha)$ is independent of parametrization.

As the appropriate physical interpretation of $L(\alpha)$ we take

Assumption #8 (The Clock Hypothesis): If $\alpha\colon [a,b] \to \mathcal{M}$ is the worldline of some material particle in $\mathcal{M}$, then $L(\alpha)$ is interpreted as the time lapse between $\alpha(a)$ and $\alpha(b)$ as measured by an ideal standard clock which is carried along by the particle.

The motivation for Assumption #8 is at the same time "obvious" and subtle. For it we shall require a special case of a result proved in a much more general context later (see Lemma 4.5.3).

Theorem 1.5.6. Let $p,q \in \mathcal{M}$. Then $p \ll q$ if and only if there exists a smooth, future-directed timelike curve $\alpha\colon [a,b] \to \mathcal{M}$ such that $\alpha(a) = p$ and $\alpha(b) = q$.

Now, in order to motivate the Clock Hypothesis let us partition $[a,b]$ into subintervals by $a = t_0 < t_1 < \ldots < t_{n-1} < t_n = b$. Then, by Theorem 1.5.6, $\alpha(a) = \alpha(t_0) \ll \alpha(t_1) \ll \ldots \ll \alpha(t_{n-1}) \ll \alpha(t_n) = \alpha(b)$ so each of the displacement vectors $x_i = \alpha(t_i) - \alpha(t_{i-1})$ is timelike and future-directed. $\tau(x_i)$ is then interpreted as the time lapse between $\alpha(t_i)$ and $\alpha(t_{i-1})$ as measured by an admissible observer who is present at both events. If the "material particle" whose worldline is represented by α has constant velocity between the events $\alpha(t_{i-1})$ and $\alpha(t_i)$, then $\tau(x_i)$ would be the time lapse between these events as measured by a clock carried along by the particle. Relative to any admissible frame,

$$\tau(x_i) = \sqrt{-\eta_{ab} \Delta x_i^a \Delta x_i^b} = \sqrt{-\eta_{ab} \frac{\Delta x_i^a}{\Delta t_i} \frac{\Delta x_i^b}{\Delta t_i}}\ \Delta t_i$$

By choosing Δt_i sufficiently small, Δx_i^4 can be made small (by continuity of α) and, since the velocity of the particle relative to our frame of reference is nearly constant over small x^4-time intervals, $\tau(x_i)$ should be a good approximation to the time lapse between $\alpha(t_{i-1})$ and $\alpha(t_i)$ measured by the material particle. Consequently, the sum

$$\sum_{i=1}^{n} \sqrt{-\eta_{ab} \frac{\Delta x_i^a}{\Delta t_i} \frac{\Delta x_i^b}{\Delta t_i}}\ \Delta t_i \qquad (73)$$

approximates the time lapse between $\alpha(a)$ and $\alpha(b)$ that this particle measures. The approximations become better as the Δt_i approach 0 and, in the limit, the sum (73) approaches the integral in the definition of $L(\alpha)$.

The argument seems persuasive enough, but it clearly rests on an assumption about the behavior of ideal clocks that we did not make in section 1.2, namely, that acceleration as such has no effect on their rates, i.e., that the "instantaneous rate" of such a clock depends only on its instantaneous speed and not the rate at which this speed is changing. Justifying such an assumption is a nontrivial matter. One must perform experiments with various types of "clocks" subjected to real accelerations and, in the end, will probably be forced to a more modest proposal ("The Clock Hypothesis is valid for such and such a clock over such and such a range of accelerations."). Indeed, one need only ponder the effects of the rapid deceleration experienced by a wrist watch when it is dropped to the floor.

Proper time length along the worldline of a material particle leads us to the most useful reparametrization of such a curve. First let us appeal to Exercise 1.5.9 and translate the domain of our curve α in the real line if necessary and assume $0 \in I$. Now define the *proper time function* $\tau(t)$ on I by

$$\tau = \tau(t) = \int_0^t |\alpha'(t) \cdot \alpha'(t)|^{1/2} \, dt \ .$$

Thus, $\dfrac{d\tau}{dt} = |\alpha'(t) \cdot \alpha'(t)|^{1/2}$ which is positive and C^∞ since $\alpha'(t)$ is timelike. The inverse $t = h(\tau)$ therefore satisfies $\dfrac{dh}{d\tau} = (d\tau/dt)^{-1} > 0$ and we conclude that τ is an acceptable parameter for α (we are simply parametrizing the worldline α by the time readings actually recorded along α). We shall abuse our notation somewhat and use the same names for the coordinate functions of α relative to an admissible basis:

$$\alpha(\tau) = (x^1(\tau), x^2(\tau), x^3(\tau), x^4(\tau)) \tag{74}$$

The reparametrization (74) of α will be called its *proper time parametrization* and is entirely analogous to the arc length parametrization of a curve in $\mathbb{R}^3$ except that it is defined only for timelike curves. In particular, when parametrized by τ, α has "unit speed":

Exercise 1.5.10. Show that if $\alpha = \alpha(\tau)$ is a smooth, timelike curve parametrized by proper time, then $\alpha'(\tau) \cdot \alpha'(\tau) = -1$.

Exercise 1.5.11. Let $\alpha = \alpha(t) = x_0 + t(x - x_0)$, where $Q(x - x_0) < 0$ and $t \in I$, be the worldline of a free material particle. Show that $\tau = \tau(x - x_0)t$.

The Riemann sum argument by which we motivated Assumption #8 has another use as well. Given a smooth, future-directed timelike curve α: $[a,b] \to \mathcal{M}$ from $\alpha(a) = q$ to $\alpha(b) = p$, the integral defining $L(\alpha)$ can be arbitrarily well approximated by a sum of the form (73), i.e., by

$$\tau(\alpha(t_1) - \alpha(t_0)) + \tau(\alpha(t_2) - \alpha(t_1)) + \ldots + \tau(\alpha(t_n) - \alpha(t_{n-1})) \ .$$

But, by the Reversed Triangle Inequality, this sum is at most

$$\tau(\alpha(t_1) - \alpha(t_0) + \alpha(t_2) - \alpha(t_1) + \ldots + \alpha(t_n) - \alpha(t_{n-1})) = \tau(p - q) \ .$$

Thus, in the limit one would expect to conclude that $\tau(p - q) \geq L(\alpha)$, i.e., that the time lapse between p and q is longest for the observer at rest (in an admissible frame) who experiences both (again, "moving clocks run slow"). Thus, we are led to conjecture the following result which will be proved in a much more general context later (*Lemma* 4.5.8).

Theorem 1.5.7. Let α : $[a,b] \to \mathcal{M}$ be a smooth, future-directed timelike curve in $\mathcal{M}$ from $\alpha(a) = q$ to $\alpha(b) = p$. Then

$$L(\alpha) \leq \tau(p - q) \tag{74}$$

and equality holds if and only if α is the worldline of a free material particle.

1.6. Spacelike Vectors

Now we turn to spacelike separations, i.e., we consider two events x_0 and x for which $Q(x - x_0) > 0$. Relative to any admissible basis we have $(\Delta x^1)^2 + (\Delta x^2)^2 + (\Delta x^3)^2 > (\Delta x^4)^2$ so that $x - x_0$ lies *outside* the null cone at x_0 and there is obviously no admissible basis in which the spatial separation of the two events is zero, i.e., there is no admissible observer who can experience both events (to do so he would have to travel faster than the speed of light). However, an argument analogous to that given at the beginning of section 1.5 will show that there is a frame in which x_0 and x are simultaneous.

Exercise 1.6.1. Show that if $Q(x - x_0) > 0$, then there is an admissible basis $\{\bar{e}_a\}$ for $\mathcal{M}$ relative to which $\Delta\bar{x}^4 = 0$. *Hint.* With $\{e_a\}$ arbitrary, take $\beta_r = \Delta x^4 / \Delta\vec{x}$ and $d^i = \Delta x^i / \Delta\vec{x}$.

Exercise 1.6.2. Show that if $Q(x - x_0) > 0$ and s is an arbitrary real number (positive, negative or zero), then there is an admissible basis for $\mathcal{M}$ relative to which the temporal separation of x_0 and x is s (so that admissible observers will, in general, not even agree on the *temporal order* of x_0 and x).

Since $((\Delta x^1)^2 + (\Delta x^2)^2 + (\Delta x^3)^2)^{1/2} = \sqrt{(\Delta x^4)^2 + Q(x - x_0)}$ in any admissible frame and since $(\Delta x^4)^2$ can assume any non-negative real value, the spatial separation of x_0 and x can assume any value greater than or equal to $\sqrt{Q(x - x_0)}$; there is no frame in which the spatial separation is less than this value. For any two events x_0 and x for which $Q(x - x_0) > 0$ we define the *proper spatial separation* of x_0 and x by

$$S(x - x_0) = \sqrt{Q(x - x_0)}$$

and regard it as the spatial separation of x_0 and x in any frame in which x_0 and x are simultaneous.

Let T be an arbitrary timelike straight line containing x_0 (we have seen that T can be identified with the worldline of some observer at rest in an admissible frame of reference, but not necessarily stationed at the origin of the spatial coordinate system of this frame - we consider the special case of a time axis shortly). Let $x \in \mathcal{M}$ be such that $x - x_0$ is spacelike and let x_1 and x_2 be the points of intersection of T with $C_N(x)$ as shown in Figure 1.8. We claim that

$$S^2(x - x_0) = \tau(x_0 - x_1)\,\tau(x_2 - x_0) \ . \tag{75}$$

Figure 1.4

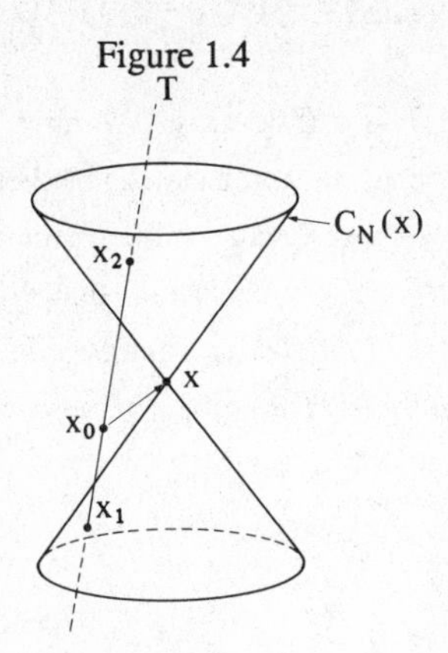

To prove (75) we observer that, since $x - x_1$ is null,

$$0 = Q(x - x_1) = Q((x_0 - x_1) + (x - x_0))$$

$$0 = -\tau^2(x_0 - x_1) + 2(x_0 - x_1) \bullet (x - x_0) + S^2(x - x_0) \ . \tag{76}$$

Similarly, since $x_2 - x$ is null,

$$0 = -\tau^2(x_2 - x_0) - 2(x_2 - x_0) \bullet (x - x_0) + S^2(x - x_0) \ . \tag{77}$$

Now, there exists a constant $k > 0$ such that $x_2 - x_0 = k(x_0 - x_1)$ so $\tau^2(x_2 - x_0) = k^2\tau^2(x_0 - x_1)$. Multiplying (76) by k and adding the result to (77) therefore yields

$$-(k + k^2)\tau^2(x_0 - x_1) + (k + 1)S^2(x - x_0) = 0 \ .$$

Since $k + 1 \neq 0$ this can be written

$$\begin{aligned} S^2(x - x_0) &= k\tau^2(x_0 - x_1) \\ &= \tau(x_0 - x_1)(k\tau(x_0 - x_1)) \\ &= \tau(x_0 - x_1)\tau(x_2 - x_0) \end{aligned}$$

as required.

Remarks. Suppose the spacelike displacement vector $x - x_0$ is orthogonal to the timelike straight line T. Then (with the notation as above), $(x_0 - x_1) \bullet (x - x_0) = (x_2 - x_0) \bullet (x - x_0) = 0$ so (76) and (77) yield $S(x - x_0) = \tau(x_0 - x_1) = \tau(x_2 - x_0)$ which we prefer to write as

$$S(x - x_0) = \frac{1}{2}(\tau(x_0 - x_1) + \tau(x_2 - x_0)) \ . \tag{78}$$

In particular, this is true if T is a time axis. We have seen that, in this case, T can be identified with the worldline of an admissible observer O and the events x_0 and x are simultaneous in this observer's reference frame. But then $S(x - x_0)$ is the O-distance between x_0 and x. Since x_0 lies on T we find that (78) admits the following physical interpretation: *the O-distance of an event **x** from an admissible observer **O** is one-half the time lapse measured by **O** between the emission and reception of light signals connecting **O** with x.*

Exercise 1.6.3 Let x, x_0 and x_1 be events for which $x - x_0$ and $x_1 - x$ are spacelike and orthogonal. Show that

$$S^2(x_1 - x_0) = S^2(x_1 - x) + S^2(x - x_0)$$

and interpret the result physically by considering a time axis T which is orthogonal to both $x - x_0$ and $x_1 - x$.

PROBLEMS

1.A. Geometrical Representation of the Special Lorentz Transformations

The special Lorentz transformation (59) corresponds to a physical situation in which the motion is one-dimensional so that two spatial coordinates can be suppressed and spacetime is essentially two-dimensional. We construct a geometrical representation for such a Lorentz transformation in the plane as follows: Label two orthogonal Cartesian coordinate axes in the plane x^1 and x^4. The $\bar{x}^1$-axis coincides with the set of events with $\bar{x}^4 = 0$, i.e., with the straight line $x^4 = \beta_r x^1$. Similarly, the $\bar{x}^4$-axis is identified with the line $x^4 = (1/\beta_r)x^1$. Since (59) preserves the Lorentz quadratic form, the hyperbolas $(x^1)^2 - (x^4)^2 = 1$ and $(x^1)^2 - (x^4)^2 = -1$ coincide with the curves $(\bar{x}^1)^2 - (\bar{x}^4)^2 = 1$ and $(\bar{x}^1)^2 - (\bar{x}^4)^2 = -1$ respectively. Thus, the length scales on the barred axes must differ from those on the unbarred axes.

Figure 1.5

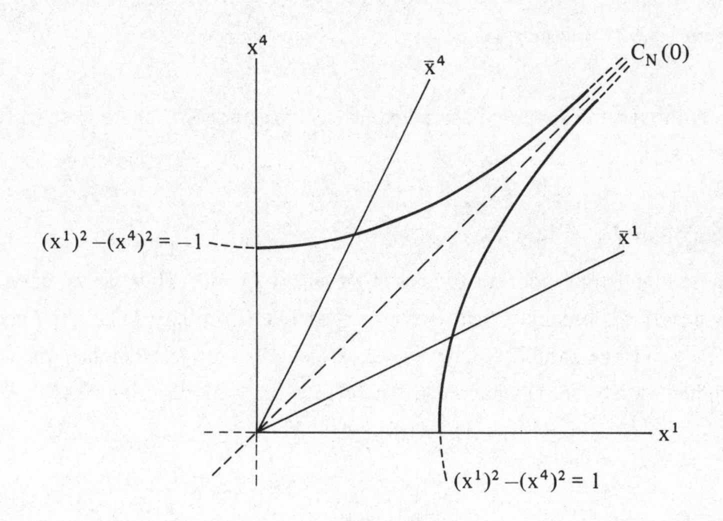

1. Show that one unit of length on the $\bar{x}^1$ - and $\bar{x}^4$-axes has Euclidean length $(1+\beta_r^2)^{1/2}(1-\beta_r^2)^{-1/2}$ in the picture.
2. Show that each of the hyperbolas $(x^1)^2 - (x^4)^2 = \pm k\,(k>0)$ intersect the $\bar{x}^1$ - and $\bar{x}^4$-axes at points which are Euclidean distance $(1+\beta_r^2)^{1/2}(1-\beta_r^2)^{-1/2}k$ from the origin.
3. Show that, with these calibrations of the barred axes, both sets of coordinates for any event can be obtained geometrically by projecting parallel to the opposite axes.
4. Describe geometrical representations of the relativity of simultaneity, time dilation and length contraction.

1.B. The Twin "Paradox"

Special relativity is plagued (blessed?) with dozens of "thought experiments" which purport to uncover inconsistencies in the theory. Some are instructive (like Exercise 1.3.13); some are just stupid. Here's an example of the latter variety. Suppose that, at (0,0,0,0), two identical twins part company. One remains at rest in the admissible frame in which he was born. The other is transported away at some constant speed to a distant point in space where he turns around and returns at the same constant speed to rejoin his brother. At the reunion the stationary twin finds that he is considerably older than his more adventurous brother. Not surprising; after all, "moving clocks run slow". However, is it not true that, from the point-of-view of the "rocket twin" it is the "stationary" brother who has been moving and must therefore be the younger of the two?

Resolve the difficulty. *Hint*. Draw the worldlines of the two brothers. Which is appropriate, Theorem 1.5.5 or Theorem 1.5.6?

Remark. Although the supposed "paradox" is easily disposed of, there is actually more to this (see [Per]).

1.C. Orthogonality

Suppose x and y are nonzero vectors in $\mathcal{M}$ with $x \cdot y = 0$. Thus far we have shown the following: If x is timelike, then y must be spacelike (Corollary 1.3.2). If x and y are null, then they must be parallel (Lemma 1.4.1). If x and y are spacelike, then their proper spatial lengths satisfy the Pythagorean Theorem $S^2(x+y) = S^2(x) + S^2(y)$ (Exercise 1.6.3).

1. Can a spacelike vector be orthogonal to a nonzero null vector?

2. Show that if x is timelike and y is spacelike, then their spacetime directions are "conjugate with respect to the null cone" in the sense of the following diagram. *Hint.* Assume without loss of generality that $\tau(x) = S(y)$.

Figure 1.6

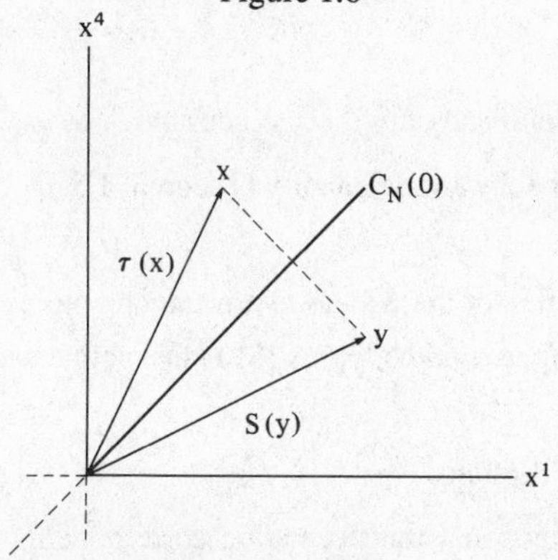

1.D. Image of a Spherical Electromagnetic Wavefront

Let (x^1, x^2, x^3, x^4) and $(\bar{x}^1, \bar{x}^2, \bar{x}^3, \bar{x}^4)$ be two admissible coordinate systems for $\mathcal{M}$ related by the special Lorentz transformation (59). The upper null cone $C_N^+(0)$ is described in these two coordinate systems by $x^4 = \rho$ and $\bar{x}^4 = \bar{\rho}$, where $\rho = ((x^1)^2 + (x^2)^2 + (x^3)^2)^{1/2}$ and similarly for $\bar{\rho}$. Intersect $C_N^+(0)$ with a hyperplane $x^4 =$ a positive constant and show that in the unbarred coordinate system, the points on this intersection satisfy

$$\rho = \bar{\rho} \, / \, (\cosh \theta_r - \sinh \theta_r \cos \phi)$$

where $\cos \phi = x^1/\rho$. Conclude that the projection of this intersection onto a hyperplane of constant x^4 is an ellipsoid of revolution with major axis along the x^1-axis and eccentricity $\beta_r = \tanh \theta_r$. Interpret this fact physically.

1.E. The Distance Between Timelike Straight Lines

Let x_0 and x'_0 be two events and A and A' two timelike unit vectors ($A \cdot A = A' \cdot A' = -1$). We consider the two timelike straight lines $L = \{x_0 + \tau A : \tau \in \mathbb{R}\}$ and $L' = \{x'_0 + \tau' A' : \tau' \in \mathbb{R}\}$, each parametrized by proper time. Fix an arbitrary point $q = x_0 + \tau A$ on L and let $p = x'_0 + \tau' A'$ be the unique point on L' which is simultaneous with q in an admissible frame whose time axis is parallel to L. Finally let $N = p - q$ be the displacement vector from q to p.

1. N is orthogonal to A so $\tau' = (A \cdot A')^{-1}(\tau + A \cdot (x'_0 - x_0))$.

2. $N = -\tau(A + (A \cdot A')^{-1}A') + C$, where $C = (x'_0 - x_0) - \left(\frac{A \cdot (x'_0 - x_0)}{A \cdot A'}\right)A'$.

3. The *distance* D between L and L' measured at q is defined by $D = S(N)$. Show that

$$D^2 = \tau^2(1 - (A \cdot A')^{-2}) - 2\tau C \cdot (A + (A \cdot A')^{-1}A') + C \cdot C \ .$$

4. L and L' are parallel if and only if D^2 is constant.

Hint. Use the Reversed Schwartz Inequality (Theorem 1.5.4).

Remark. A generalization of these ideas to arbitrary timelike curves was used by Born in his relativistic theory of the rigid body; see [Sy1] for more details.

1.F. Invariant Null Directions

Show that the orthogonal transformation corresponding to any proper, orthochronous Lorentz transformation leaves invariant at least one light ray. *Hint.* Use the Brouwer Fixed Point Theorem: Any continuous map of a closed ball in $\mathbb{R}^3$ into itself leaves at least one point fixed.

1.G. Infinitesimal Matrices and Commutation Relations

Let $R_3(t)$ be the rotation in $\mathcal{L}$ defined by

$$R_3(t) = \begin{bmatrix} \cos t & -\sin t & 0 & 0 \\ \sin t & \cos t & 0 & 0 \\ 0 & 0 & 1 & 0 \\ 0 & 0 & 0 & 1 \end{bmatrix}$$

1. Expand $\cos t$ and $\sin t$ into power series about $t = 0$ and show that, for small t, $R_3(t)$ can be approximated by $I_{4\times 4} + R_3 t$, where $I_{4\times 4}$ is the 4×4 identity matrix and R_3 is the *infinitesimal matrix corresponding to rotation about the* $\mathbf{x}^3$*-axis* defined by

$$R_3 = \begin{bmatrix} 0 & -1 & 0 & 0 \\ 1 & 0 & 0 & 0 \\ 0 & 0 & 0 & 0 \\ 0 & 0 & 0 & 0 \end{bmatrix} .$$

Observe that $R_3 = \left.\frac{dR_3(t)}{dt}\right|_{t=0}$. Heuristically, one thinks of $I_{4\times 4} + R_3 t$ as the matrix of an "infinitely small rotation" about the x^3-axis.

2. Define analogous rotations $R_1(t)$ and $R_2(t)$ about the x^1- and x^2-axes and corresponding infinitesimal matrices R_1 and R_2.

3. Let $[A,B] = AB - BA$ be the commutator of the two matrices A and B. Prove the following *commutation relations:*

$$[R_1,R_2] = R_3$$

$$[R_2,R_3] = R_1$$

$$[R_3,R_1] = R_2 \ .$$

4. Show that, for small θ_r, the special Lorentz transformation $L(\theta_r)$ can be approximated by $I_{4\times 4} + L\,\theta_r$, where

$$L = \left.\frac{dL\,(\theta_r)}{d\,\theta_r}\right|_{\theta_r=0} = \begin{bmatrix} 0 & 0 & 0 & -1 \\ 0 & 0 & 0 & 0 \\ 0 & 0 & 0 & 0 \\ -1 & 0 & 0 & 0 \end{bmatrix}$$

is the *infinitesimal matrix corresponding to a Lorentz "boost" in the* $\mathbf{x}^1$*-direction.*

5. Prove the following commutation relations:

$[R_1,L] = 0$

$[R_2,L] =$ - the infinitesimal matrix for the Lorentz boost in the x^3-direction

$[R_3,L] =$ - the infinitesimal matrix for the Lorentz boost in the x^2-direction.

1.H. The Spinor Map

We let $SL(2,C)$ denote the group of complex 2×2 matrices with determinant 1. Complex conjugation will be indicated by * and the conjugate transpose of a matrix $\mathcal{A}$ will be denoted $\mathcal{A}^+$. A Hermitian matrix is one which equals its conjugate transpose.

1. Fix an $\mathcal{A} = \begin{bmatrix} a & b \\ c & d \end{bmatrix}$ in $SL(2,C)$. Show that $\mathcal{A}$ gives rise to a determinant preserving map of the set of 2×2 Hermitian matrices to itself defined by $Q \to \overline{Q} = \mathcal{A}\,Q\,\mathcal{A}^+$ for every Hermitian Q.

2. Every 2×2 Hermitian Q can be written in the form $\begin{bmatrix} x^4 + x^3 & x^1 + ix^2 \\ x^1 - ix^2 & x^4 - x^3 \end{bmatrix}$ where the x^a, $a = 1,2,3,4$, are real, i.e., every such Q is a real linear combination of σ_a, $a = 1,2,3,4$, where

$$\sigma_1 = \begin{bmatrix} 0 & 1 \\ 1 & 0 \end{bmatrix}, \quad \sigma_2 = \begin{bmatrix} 0 & i \\ -i & 0 \end{bmatrix}, \quad \sigma_3 = \begin{bmatrix} 1 & 0 \\ 0 & -1 \end{bmatrix}$$

are the *Pauli spin matrices* and σ_4 is the 2×2 identity matrix.

3. Let $Q = x^a \sigma_a$ as in #2 and, similarly, $\bar{Q} = \mathcal{A} Q \mathcal{A}^+ = \bar{x}^a \sigma_a$. Write out the map $Q \to \bar{Q}$ explicitly to show that

$$\begin{bmatrix} \bar{x}^1 \\ \bar{x}^2 \\ \bar{x}^3 \\ \bar{x}^4 \end{bmatrix} = \begin{bmatrix} \mathrm{Re}(ad^* + bc^*) & -\mathrm{Im}(ad^* - bc^*) & \mathrm{Re}(ac^* - bd^*) & \mathrm{Re}(ac^* + bd^*) \\ \mathrm{Im}(ad^* + bc^*) & \mathrm{Re}(ad^* - bc^*) & \mathrm{Im}(ac^* - bd^*) & \mathrm{Im}(ac^* + bd^*) \\ \mathrm{Re}(ab^* - cd^*) & -\mathrm{Im}(ab^* - cd^*) & 1/2(aa^* - bb^* - cc^* + dd^*) & 1/2(aa^* + bb^* - cc^* - dd^*) \\ \mathrm{Re}(ab^* + cd^*) & -\mathrm{Im}(ab^* + cd^*) & 1/2(aa^* - bb^* + cc^* - dd^*) & 1/2(aa^* + bb^* + cc^* + dd^*) \end{bmatrix} \begin{bmatrix} x^1 \\ x^2 \\ x^3 \\ x^4 \end{bmatrix}$$

4. Show that $\eta_{ab} \bar{x}^a \bar{x}^b = \eta_{ab} x^a x^b$ so that the matrix in #3 is a general Lorentz transformation. Show also that this Lorentz transformation is orthochronous and proper. *Hint.* To show that the transformation is proper use the fact that $SL(2,C)$ is a connected subspace of $C^4 = \mathbb{R}^8$ and that the determinant of the matrix in #3 is a continuous function of a,b,c and d. This argument requires some familiarity with point-set topology.

5. Denoting the matrix in #3 by $[\Lambda^a_b]$ show that

$$\Lambda^a_b = \frac{1}{2} \mathrm{tr}\,(\sigma_a \, \mathcal{A} \sigma_b \, \mathcal{A}^+) .$$

Hint. First show that, for any real numbers $\bar{x}^a$, $a = 1,2,3,4$, $\bar{x}^a = \frac{1}{2} \mathrm{tr}\,(\sigma_a \sigma_b \bar{x}^b)$.

6. The map from $SL(2,C)$ to $\mathcal{L}$ defined by $\mathcal{A} \to L_{\mathcal{A}}$, where $L_{\mathcal{A}}$ is the matrix in #3 is called the *spinor map*. Show that, for any Hermitian Q and any two elements $\mathcal{A}_1$ and $\mathcal{A}_2$ of $SL(2,C)$, $(\mathcal{A}_2 \mathcal{A}_1)\, Q\, (\mathcal{A}_2 \mathcal{A}_1)^+ = \mathcal{A}_2 (\mathcal{A}_1\, Q\, \mathcal{A}_1^+)\, \mathcal{A}_2^+$ and conclude that the spinor map is a group homomorphism, i.e., preserves matrix products.

7. The spinor map carries both $\mathcal{A}$ and $-\mathcal{A}$ onto the same Lorentz transformation.

8. The spinor map is precisely two-to-one, i.e., $L_{\mathcal{A}_1} = L_{\mathcal{A}_2}$ implies $\mathcal{A}_1 = \pm\mathcal{A}_2$. *Hint.* Consider $\mathcal{A} = \mathcal{A}_1\ \mathcal{A}_2^{-1}$.

9. The spinor map carries the unitary elements of $SL(2,C)$ onto rotations in $\mathcal{L}$. *Hint.* $\mathcal{A}$ is unitary if $\mathcal{A}^{-1} = \mathcal{A}^{+}$.

10. Prove that the converse of #9 is also true, i.e., that every rotation in $\mathcal{L}$ is the image under the spinor map of some unitary element of $SL(2,C)$.

11. Let θ_r be a real number and $L(\theta_r)$ the corresponding special Lorentz transformation. Show that $L(\theta_r)$ is the image under the spinor map of $\pm\hat{L}(\theta_r) \in SL(2,C)$, where

$$\hat{L}(\theta_r) = \begin{bmatrix} \cosh(\theta_r/2) & -\sinh(\theta_r/2) \\ -\sinh(\theta_r/2) & \cosh(\theta_r/2) \end{bmatrix}$$

Hint. Use the following identities: $\cosh^2 x - \sinh^2 x = 1$, $\cosh 2x = \cosh^2 x + \sinh^2 x$, $\sinh 2x = 2\sinh x\ \cosh x$.

12. Show that the spinor map is surjective. *Hint.* Use Theorem 1.3.5 and the results of #6 and #11.

Summary. The spinor map $\mathcal{A} \to L_{\mathcal{A}}$ is a two-to-one homomorphism of $SL(2,C)$ onto $\mathcal{L}$ which carries $\pm\mathcal{A}$ onto $L_{\mathcal{A}}$ for every $\mathcal{A}$ in $SL(2,C)$ and which, when restricted to the unitary subgroup of $SL(2,C)$, maps onto the rotation subgroup of $\mathcal{L}$.

1.I. The Zeeman Topology for $\mathcal{M}$

This problem requires some familiarity with elementary point-set topology. We let $\mathcal{M}^E$ denote Minkowski spacetime with the usual Euclidean topology of $\mathbb{R}^4$. The (subspace) topology induced by $\mathcal{M}^E$ on any spacelike hyperplane or timelike straight line is obviously the Euclidean topology. The *Zeeman topology* for $\mathcal{M}$ is the finest topology with this property, i.e., it is the topology in which a subset U of $\mathcal{M}$ is open if and only if $U \cap A$ is a Euclidean open subset of A for every spacelike hyperplane and timelike straight line A. Endowed with this topology we denote $\mathcal{M}$ by $\mathcal{M}^Z$.

1. Every $\mathcal{M}^E$-open set is open in $\mathcal{M}^Z$, but the converse is not true. *Hint.* For each $\varepsilon > 0$, let $N_\varepsilon^E(x)$ be the usual Euclidean open ε-ball about x. Define the *Z-open ball* $N_\varepsilon^Z(x)$ by $N_\varepsilon^Z(x) = (N^E(x) - C_N(x)) \cup \{x\}$ (throw away the null cone at x and replace the point x). $N_\varepsilon^Z(x)$ is Z-open, but not E-open.

2. The subspace topology induced by $\mathcal{M}^Z$ on any light ray is discrete.

3. $\mathcal{M}^Z$ is Hausdorff, but not normal. *Hint.* To prove non-normality use the Baire Category Theorem.

4. A sequence $X = \{x_n\}_{n=1}^{\infty}$ of distinct points in $\mathcal{M}$ which converges to a point x in $\mathcal{M}$ in the Euclidean topology, but does not converge in $\mathcal{M}^Z$ is called a *Zeno sequence.*

   (a) A Zeno sequence is closed and discrete in $\mathcal{M}^Z$.

   (b) Fix an x in $\mathcal{M}$. Let $\{L_n\}_{n=1}^{\infty}$ be a sequence of (not necessarily distinct) light rays through x. Select x_n in $L_n - \{x\}$ such that $d(x,x_n) \to 0$ as $n \to \infty$, where d is the Euclidean metric. Then $\{x_n\}_{n=1}^{\infty}$ is a Zeno sequence in $\mathcal{M}$.

   (c) Fix an x in $\mathcal{M}$. Let $\{L_n\}_{n=1}^{\infty}$ be a sequence of distinct timelike straight lines through x. Construct a sequence as in (b) and prove that it is a Zeno sequence. *Hint.* Show that any spacelike hyperplane or timelike straight line contains at most finitely many terms of the sequence.

   (d) Repeat the construction in (b) and (c) with a sequence of distinct spacelike straight lines such that at most finitely many lie in any given spacelike hyperplane.

5. The Z-open balls $N_\varepsilon^Z(x)$, $\varepsilon > 0$, do *not* form a local base at x.

6. $\mathcal{M}^Z$ is not locally compact; indeed, no point of $\mathcal{M}^Z$ has a compact neighborhood.

7. Let $f: [0,1] \to \mathcal{M}^Z$ be a continuous map of the unit interval into $\mathcal{M}^Z$ which is strictly order preserving, i.e., $t_1 < t_2$ in $[0,1]$ implies $f(t_1) \ll f(t_2)$ in $\mathcal{M}$. Let $x = f(0)$.

   (a) There exists an $\varepsilon > 0$ such that $f[0,\varepsilon]$ lies along a timelike straight line through x. *Hint.* Suppose not and inductively construct a sequence $\{x_n\}_{n=1}^{\infty}$ of points lying on distinct timelike straight lines through x and with $f^{-1}(x_n)$ in $(0,1/n)$. Now use #4(c) and #4(a).

   (b) $f[0,1]$ is piecewise linear, consisting of a finite number of intervals along timelike straight lines.

   (c) There exists a homeomorphism g of $[0,1]$ onto itself such that $f \circ g: [0,1] \to \mathcal{M}^Z$ is a piecewise linear embedding.

Remarks. Problem #7 indicates a rather close connection between the linear structure of $\mathcal{M}$ and the topological structure of $\mathcal{M}^Z$ (the corresponding results for $\mathcal{M}^E$ are obviously false). Zeeman [Z2] pursued these matters in more detail and proved the quite remarkable fact that any homeomorphism of $\mathcal{M}^{Z}$ onto itself is either an orthogonal transformation of $\mathcal{M}$, a translation, a (nonzero) scalar multiplication or a composition of these (and therefore is, in particular, linear). The ideas have been generalized by Göbel [G] and Hawking, King and McCarthy [HKM].

CHAPTER 2

SOME CONCEPTS FROM RELATIVISTIC MECHANICS

2.1. Material Particles and Photons

A *material particle* in $\mathcal{M}$ is a pair (α, m), where $\alpha: I \to \mathcal{M}$ is the worldline of a material particle and m is a positive real number called the particle's *proper mass*. A *free material particle* is defined in the same way, but with α the worldline of a free material particle. If α is parametrized by proper time τ, then $\alpha'(\tau) \bullet \alpha'(\tau) = -1$ and we shall refer to $\alpha'(\tau)$ as the particle's *4-velocity* (generally denoted U). It is sometimes convenient to include the particle's proper mass m in the parametrization by defining $\sigma = \tau/m$. Then $\alpha'(\sigma) \bullet \alpha'(\sigma) = -m^2$ and we shall call $\alpha'(\sigma) = m\alpha'(\tau)$ the particle's *4-momentum* (designated P). Indeed, one could define a material particle as a smooth, future-directed, timelike curve $\alpha = \alpha(t)$ for which $\alpha'(t) \bullet \alpha'(t)$ is constant then take $\tau(\alpha'(t))$ to be the mass of the particle.

Relative to any admissible coordinate system (x^1, x^2, x^3, x^4) the 4-velocity $\alpha'(\tau)$ is given by $\left(\frac{dx^1}{d\tau}, \frac{dx^2}{d\tau}, \frac{dx^3}{d\tau}, \frac{dx^4}{d\tau}\right)$. If $\left(\frac{d\bar{x}^1}{d\tau}, \frac{d\bar{x}^2}{d\tau}, \frac{d\bar{x}^3}{d\tau}, \frac{d\bar{x}^4}{d\tau}\right)$ are the components of $\alpha'(\tau)$ in another system, related to (x^a) by $\bar{x}^a = \Lambda^a_b x^b$, then $\frac{d\bar{x}^a}{d\tau} = \Lambda^a_b \frac{dx^b}{d\tau}$, i.e., $\bar{U}^a = \Lambda^a_b U^b$, $a = 1,2,3,4$. The components of (α, m)'s 4-momentum satisfy the same transformation law. In general, any "object" which is described in each admissible coordinatization of $\mathcal{M}$ by four numbers (components) (V^1, V^2, V^3, V^4), $(\bar{V}^1, \bar{V}^2, \bar{V}^3, \bar{V}^4)$, ... related by the transformation law

$$\bar{V}^a = \Lambda^a_b V^b, \ a = 1,2,3,4, \tag{1}$$

is called a (*contravariant*) *4-vector* on $\mathcal{M}$. Since these objects "transform like points" they constitute a natural generalization to $\mathcal{M}$ of the familiar notion of a "vector" in $\mathbb{R}^3$. Another important example is the *4-acceleration* $\alpha''(\tau)$ of (α, m) defined by $\alpha''(\tau) = \left(\frac{d^2x^1}{d\tau^2}, \frac{d^2x^2}{d\tau^2}, \frac{d^2x^3}{d\tau^2}, \frac{d^2x^4}{d\tau^2}\right)$ and designated A.

Exercise 2.1.1. Show that $U \bullet A = 0$ so that the 4-acceleration of (α, m) is either zero or spacelike. *Hint.* $\alpha'(\tau) \bullet \alpha'(\tau) = -1$ for all τ.

An admissible observer is more likely to parametrize a worldline by his own time coordinate x^4. Then $\alpha'(x^4) = \left(\frac{dx^1}{dx^4}, \frac{dx^2}{dx^4}, \frac{dx^3}{dx^4}, 1 \right)$ so $\alpha'(x^4) \bullet \alpha'(x^4) = 1 - \beta^2(x^4)$, where $\beta = \beta(x^4) = ((dx^1/dx^4)^2 + (dx^2/dx^4)^2 + (dx^3/dx^4)^2)^{1/2}$ is the usual magnitude of the particle's instantaneous velocity 3-vector in the given frame, i.e., its speed in that frame. Thus,

$$\tau = \int_0^{x^4} \sqrt{1 - \beta^2}\, dx^4 \ .$$

Moreover, the particle's 4-velocity and 4-momentum are now given by $U = \left(\frac{dx^1}{d\tau}, \frac{dx^2}{d\tau}, \frac{dx^3}{d\tau}, \frac{dx^4}{d\tau} \right) = \left(\frac{dx^1}{dx^4}, \frac{dx^2}{dx^4}, \frac{dx^3}{dx^4}, \frac{dx^4}{dx^4} \right) \frac{dx^4}{d\tau}$, i.e.,

$$U = (1 - \beta^2)^{-1/2} \left(\frac{dx^1}{dx^4}, \frac{dx^2}{dx^4}, \frac{dx^3}{dx^4}, 1 \right) \tag{2}$$

and

$$P = m (1 - \beta^2)^{-1/2} \left(\frac{dx^1}{dx^4}, \frac{dx^2}{dx^4}, \frac{dx^3}{dx^4}, 1 \right) . \tag{3}$$

Letting $\gamma = (1 - \beta^2)^{-1/2}$, $v^i = dx^i/dx^4$ and denoting the velocity 3-vector of the particle in this frame by $\vec{v} = (v^1, v^2, v^3)$, one can write

$$U = \gamma (\vec{v}, 1) \tag{4}$$

and

$$P = m\gamma (\vec{v}, 1) \ . \tag{5}$$

Writing out the components of the particle's 4-momentum in more detail and expanding $\gamma = (1 - \beta^2)^{-1/2}$ by the Binomial Theorem leads to

$$p^i = m\, \gamma v^i = \frac{m}{\sqrt{1 - \beta^2}}\, v^i = m v^i + \frac{1}{2} m v^i \beta^2 + \dots,\ i = 1,2,3 \tag{6}$$

and

$$p^4 = m\,\gamma = \frac{m}{\sqrt{1-\beta^2}} = m + \frac{1}{2}m\beta^2 + \ldots \tag{7}$$

Identifying m with the classical "inertial mass" of Newtonian mechanics, the expansions in (6) and (7) contain some familiar terms. The mv^i in (6) make it clear that, for small relative speeds, the p^i reduce to the components of the Newtonian momentum of the particle. The quantity $\frac{m}{\sqrt{1-\beta^2}}$ is sometimes referred to as the "relativistic mass" of our particle since it permits one to maintain a formal similarity between the Newtonian and relativistic definitions of momentum. Inertial mass was regarded in classical physics as a measure of a particle's resistance to acceleration. From the relativistic point of view this resistence must become unbounded as $\beta \to 1$ and $m\,\gamma$ certainly has this property. We prefer, however, to avoid the quite misleading attitude that "mass increases with velocity" and simply abandon the Newtonian view that momentum is a linear function of speed. Turning now to (7) we recognize the term $\frac{1}{2}m\beta^2$ as the Newtonian kinetic energy of (α, m). For this reason we shall call P^4 the *total relativistic energy* of (α, m) and often denote it E.

Remark. The concept of energy in classical physics is a rather subtle one. Many different types of energy are defined in different situations, but each is in one way or another intuitively related to a system's "ability to do work". Now, simply calling P^4 the total relativistic energy of (α, m) does not insure that this intuitive interpretation is still valid in the new situation. Whether or not the name is appropriate can only be determined experimentally. In particular, one should determine whether or not the presence of the term "m" in the expansion of P^4 is consistent with this interpretation. Observe that, when $\beta = 0$ (i.e., in the "instantaneous rest frame" of the particle), $P^4 = E = m$ ($= mc^2$ in traditional units), which we interpret as saying that, even when the particle is at rest relative to an admissible frame, it still has "energy" in this frame, the amount being numerically equal to m. If this is really "energy" in the classical sense, it should be capable of doing work, i.e., it should be possible to "liberate" (and use) it. That this is indeed possible has, of course, been rather convincingly demonstrated.

We now ask the reader to show that, if one believes that "relativistic momentum" should be a 4-vector and that the spatial components defined by (6) are "right", then one has no choice about the fourth component.

Exercise 2.1.2. Show that two 4-vectors whose first three (spatial) components are the same in every admissible frame must, in fact, be equal.

Next we observe that not only material particles, but also photons possess "momentum" and "energy" and therefore should have "4-momentum". Witness, for example, the "photoelectric effect" in which photons collide with and eject electrons from their orbits in an atom. Since they propagate rectilinearly with constant speed (1) in any admissible frame, photons are in many ways analogous to free material particles. Unlike material particles, however, the photon's characteristic feature is not mass but energy (frequency, wavelength) and this feature is highly observer dependent (e.g., wavelengths of photons emitted from the atoms of a star are "red-shifted" relative to those measured on earth because the stars are receding from us due to the expansion of the universe). Moreover, there is no "proper wavelength" of a photon analogous to the "proper mass" of a free material particle since there is no admissible frame in which the photon is at rest (the term "proper wavelength" is often used to designate the wavelength measured in the rest frame of the photon's *source*, but unless the source is itself at rest in some admissible frame this varies from point to point along the source's worldline). Consequently, a "photon" in $\mathcal{M}$ must be defined somewhat differently than a free material particle. Let us consider the worldline $\alpha = \alpha(t) = x_0 + tn$, $t \in I$, of a photon, where $I \subseteq \mathbb{R}$ is an interval containing zero, x_0 is in $\mathcal{M}$ and n is a future-directed null vector.

Exercise 2.1.3. Show that, relative to any admissible basis $\{e_a\}$ for $\mathcal{M}$,

$$n = \varepsilon(\vec{d} + e_4) \ ,$$

where $\varepsilon = -n \cdot e_4$ and $\vec{d} = ((n \cdot e_1)^2 + (n \cdot e_2)^2 + (n \cdot e_3)^2)^{-1/2}((n \cdot e_1)e_1 + (n \cdot e_2)e_2 + (n \cdot e_3)e_3)$ is the *direction 3-vector* of the photon in the corresponding frame.

Now, by analogy with material particles (for which $-P \cdot e_4 = P^4 = E$) we shall call n the *4-momentum* of the photon whose worldline is α and ε the *energy* of the photon as measured in the frame $\{e_a\}$. The *frequency* ν and *wavelength* λ of the photon in this frame are given by $\nu = \varepsilon / h$ and $\lambda = 1/\nu$, where h is a constant (called *Planck's constant*).

To compare photon energies in two admissible frames we consider a second admissible basis $\{\bar{e}_a\}$ for $\mathcal{M}$. Then $n = \bar{\varepsilon}(\vec{\bar{d}} + \bar{e}_4)$, where $\bar{\varepsilon} = -n \cdot \bar{e}_4$.

Exercise 2.1.4. Show that $\bar{\varepsilon} = \gamma\varepsilon(1 - \beta_r(\vec{d} \cdot \vec{d}))$. *Hint.* Exercise 1.3.5.

But $\vec{d}$ and $\vec{d}$ lie in the subspace $[e_1, e_2, e_3]$ spanned by e_1, e_2 and e_3 and the restriction of the Lorentz inner product g to $[e_1, e_2, e_3]$ is just the usual positive definite inner product on $\mathbb{R}^3$. Thus, $\vec{d} \cdot \vec{d} = \cos\theta$, where θ is the angle in $\mathcal{S}$ between the direction of the photon and the direction of $\mathfrak{I}$. We therefore obtain

$$\frac{\bar{\varepsilon}}{\varepsilon} = \frac{\bar{\nu}}{\nu} = \frac{1 - \beta_r \cos\theta}{\sqrt{1 - \beta_r^2}} \tag{8}$$

which is the relativistic formula for the *Doppler Effect*. Using the binomial expansion for γ yields

$$\frac{\bar{\varepsilon}}{\varepsilon} = \frac{\bar{\nu}}{\nu} = (1 - \beta_r \cos\theta) + \frac{1}{2}\beta_r^2(1 - \beta_r \cos\theta) + \ldots \tag{9}$$

The first order term $1 - \beta_r \cos\theta$ is the familiar classical formula for the Doppler effect, while the remaining terms constitute the relativistic correction contributed by time dilation.

2.2. Contact Interactions

We shall henceforth use the term *free particle* to refer to either a free material particle or a photon. If $\mathcal{A}$ is a finite set of free particles, then each element of $\mathcal{A}$ has a unique 4-momentum. The sum of these 4-vectors is called the *total 4-momentum of* $\mathcal{A}$. A *contact interaction* (between free particles) in $\mathcal{M}$ is a triple $(\mathcal{A}, x, \mathcal{A}')$, where $\mathcal{A}$ and $\mathcal{A}'$ are two finite sets of free particles in $\mathcal{M}$ neither of which contains a pair of particles with linearly dependent 4-momenta and x is an event such that

(a) x is the terminal point of all the particles in $\mathcal{A}$, i.e., $x = \alpha(b)$ for all $\alpha\colon [a,b] \to \mathcal{M}$ in $\mathcal{A}$.

(b) x is the initial point of all the particles in $\mathcal{A}'$, and

(c) the total 4-momentun of $\mathcal{A}$ equals the total 4-momentum of $\mathcal{A}'$.

Remarks. x should be regarded as the "collision" of all the particles in $\mathcal{A}$, from which emerge all the particles in $\mathcal{A}'$. The requirement that neither $\mathcal{A}$ nor $\mathcal{A}'$ contain a pair of particles with linearly dependent 4-momenta is included because the particles in such a pair would presumably be physically indistinguishable. Property (c) is called the *conservation of 4-momentum* and contains the appropriate relativistic generalizations of two classical conservation principles: the conservation of momentum and the conservation of energy.

Several conclusions regarding contact interactions can be drawn directly from the results we have available. Let us consider, for example, an interaction $(\mathcal{A}, x, \mathcal{A}')$ for which $\mathcal{A}'$ consists of a single photon. Then the total 4-momentum of $\mathcal{A}'$ is null so the total 4-momentum of $\mathcal{A}$ must be null as well. Since the 4-momenta of the individual particles in $\mathcal{A}$ are all either timelike or null and future-directed, Lemma 1.5.1 implies that all of these 4-momenta must be null and parallel. Since $\mathcal{A}$ cannot contain two distinct photons with parallel 4-momenta we find that $\mathcal{A}$ must also consist of a single photon which, by (c), must have the same 4-momentum as the photon in $\mathcal{A}'$. In essence, "nothing happened at x". Our conclusion then is that *no nontrivial physical interaction of the type modelled by our definition can result in a single photon and nothing else.*

A contact interaction $(\mathcal{A}, x, \mathcal{A}')$ is called a *disintegration* or *decay* if $\mathcal{A}$ is a singleton. Suppose, for example, that $\mathcal{A}$ consists of a single material particle of proper mass m_0 and $\mathcal{A}'$ consists of two material particles of proper masses m_1 and m_2 (such disintegration do, in fact, occur in nature, e.g., in alpha-emission). Let P_0, P_1 and P_2 be the 4-momenta of the particles with proper masses m_0, m_1 and m_2 respectively. Then $P_0 = P_1 + P_2$. Appealing to the Reversed Triangle Inequality (Theorem 1.5.5) and the fact that P_1 and P_2 are linearly independent we find that

$$m_0 > m_1 + m_2 \ . \tag{10}$$

The excess mass $m_0 - (m_1 + m_2)$ of the initial particle is regarded as a measure of the amount of energy required to split m_0 into two pieces. Stated somewhat differently, when the two particles in $\mathcal{A}'$ were held together to form the single particle in $\mathcal{A}$ the "binding energy" contributes to the mass of this latter particle, while, after the decay, this difference in mass appears in the form of kinetic energy of the generated particles and released radiant energy.

Exercise 2.2.1. Analyze a disintegration interaction $(\mathcal{A}, x, \mathcal{A}')$ in which $\mathcal{A}$ consists of a single photon.

Exercise 2.2.2. Show that a free electron cannot emit or absorb a photon. **Remark.** Electrons *can* emit and absorb photons (e.g., in the photoelectric effect), but in order to do so they must be *bound* in an atom. A satisfactory description of this phenomenon requires quantum mechanics.

Problems 2.B and 2.C at the end of this chapter deal with two rather more complicated contact interactions, each of which is of great importance in elementary particle physics.

2.3. Energy-Momentum Tensors

Let us consider, informally, a "stream" of free material particles, all with the same proper mass m and parallel worldlines (i.e., traveling in the same direction with the same speed relative to an admissible frame). Each particle in the stream then has the same 4-velocity U so that one can identify a unique 4-momentum $P = mU$ associated with the stream itself. However, it is clear that this 4-vector alone does not provide all of the relevant information about the stream since, for example, it contains no clue as to the "energy density" of the stream determined by the number of particles "per unit volume". Of course, "energy density" alone is not a relativisitically meaningful (i.e., observer-independent) notion since "unit volume" is not. Indeed, what is energy density in one frame will be some combination of energy density, "energy density flux" and "momentum density flux" when viewed from another. As it happens it is quite easy to construct an "object" which contains all of this information. Denote by n_0 the number of particles per unit volume in the stream as measured in an admissible frame in which all of the particles are at rest (n_0 is called the *proper particle density* of the stream and we assume, for the present, that it is constant). Now define a 4×4 matrix $[T^{ab}]_{a,b=1,2,3,4}$ in each admissible frame by

$$T^{ab} = n_0 m U^a U^b,\ a,b = 1,2,3,4, \tag{11}$$

where (U^1, U^2, U^3, U^4) are the components of the 4-velocity vector U in this frame. In another frame with coordinates $\bar{x}^a = \Lambda^a_{\ b} x^b$ we have $\bar{U}^a = \Lambda^a_{\ b} U^b$ and so

$$\bar{T}^{ab} = \Lambda^a_{\ c} \Lambda^b_{\ d} T^{cd},\ a,b = 1,2,3,4. \tag{12}$$

Any "object" which is described in each admissible frame by a 4×4 matrix $[T^{ab}]$ and with the property that these matrices corresponding to different frames are related by the transformation law (12) is called a (*contravariant*) *4-tensor of rank* 2; the entries of each matrix are the *components* of the 4-tensor in the corresponding frame. The components of the particular 4-tensor defined by (11) admit simple physical interpretations: Using the notation of (4) and (5) and letting $n = n_0 \gamma = n_0(1-\beta^2)^{-1/2}$ we obtain:

$$T^{44} = (m\gamma)n$$

$$T^{i4} = T^{4i} = T^{44} v^i \tag{13}$$

$$= (m\,\gamma\, v^i)n$$

$$T^{ij} = T^{ji} = T^{i4} v^j$$

Exercise 2.3.1. Derive the equalities in (13).

The physical interpretation of n is clear: The proper particle density n_0 is determined by an observer for whom the stream is at rest by counting the number of particles in what he measures to be a unit volume. If another observer measures the volume occupied by this same set of particles he will find it contracted by a factor of $(1-\beta^2)^{1/2}$ so he will attribute to the stream the particle density $n_0(1-\beta^2)^{-1/2} = n$. Observing next that $m\,\gamma$ is the energy per particle measured in $\mathcal{S}$ (see (7)) and $m\,\gamma\, v^i$, $i = 1,2,3$, are the spatial components of the particle's 4-momentum in $\mathcal{S}$, we arrive at the following physical interpretations:

$$T^{44} = \text{energy density}$$

$$T^{i4} = T^{4i} = i\text{–component of energy density flux}$$

$$= i\text{–component of 4-momentum density}$$

$$T^{ij} = T^{ji} = ij\text{–momentum density flux} \tag{14}$$

$$= \text{amount of } x^i\text{–momentum that flows in the}$$

$$x^j\text{–direction per unit volume per unit time}$$

(all as measured in the frame $\mathcal{S}\,(x^1,x^2,x^3,x^4)$).

Exercise 2.3.2. Convince yourself.

The 4-tensor $[T^{ab}]$ defined by (11) is called the *energy-momentum 4-tensor* for the given particle flow. Observe that it is *symmetric,* i.e., $T^{ba} = T^{ab}$ for all $a,b = 1,2,3,4$. A "smoothed out" version of these ideas is presented in Problem 2.F on pressure free perfect fluids.

The situation we have been considering suggests a pattern that seems to persist throughout physics. By virtue of the local presence of matter, photons, electromagnetic fields, etc. a certain region of spacetime is, in some sense, "endowed with" energy and momentum which can be described at each point by a symmetric 4-tensor of rank two. In each admissible frame the entries (components) are constructed according to the prescription laid down in (14). Of course, the form of this energy-momentum 4-tensor will depend on the specific set of physical circumstances one is trying to model (a stream of material particles, a perfect fluid, a rotating mass, an electromagnetic field, etc.). The total energy-momentum 4-tensor for a given region of spacetime (which essentially describes everything there is to know about the mass-energy content of that region) will generally be a sum $T = T_1 + \ldots + T_n$, where each T_k corresponds to a particular physical field.

In order to formulate a general definition it will be convenient at this point to introduce some new terminology and notation. If V is any 4-vector with components V^a, $a = 1,2,3,4$, relative to an admissible basis $\{e_a\}$, then the numbers

$$V_a = \eta_{ab} V^b$$

are called the *covariant components* of V relative to $\{e_a\}$.

Exercise 2.3.3. Show that V_1, V_2, V_3 and V_4 are actually the components of V relative to the basis $\{e^a\}$ for $\mathcal{M}$ which is *dual* to $\{e_a\}$, defined by $e^a = \eta^{ab} e_b$. Observe also that $\{e^a\}$ is *not* an admissible basis for $\mathcal{M}$ since its time orientation is reversed.

With this one can write, for example, $V \cdot W = V^a W_a$. In addition, we can now regard a rank two 4-tensor as a real-valued, bilinear function on 4-vectors in the following way:

$$T(V,W) = T^{ab} V_a W_b \ .$$

Exercise 2.3.4. Show that this definition is independent of coordinates, i.e., that, relative to two admissible bases $\{e_a\}$ and $\{\bar{e}_a\}$, $T^{ab} V_a W_b = \bar{T}^{ab} \bar{V}_a \bar{W}_b$.

Alternatively, one can define the *covariant components* of T by $T_{ab} = \eta_{a\alpha} \eta_{b\beta} T^{\alpha\beta}$ (transformation law $\bar{T}_{ab} = \bar{\Lambda}_a^{\alpha} \bar{\Lambda}_b^{\beta} T_{\alpha\beta}$) so that $T(V,W) = T_{ab} V^a W^b$.

Such a 4-tensor T is therefore not unlike an inner product, although it need not be either symmetric or nondegenerate. In the particular case of the energy-momentum 4-tensor defined by (11) we find that

$$T(V,W) = n_0 m (U \cdot V)(U \cdot W) \tag{15}$$

and so

$$T(V,V) = n_0 m (U \cdot V)^2 \ . \tag{16}$$

If $\{e_a\}$ is any admissible basis for $\mathcal{M}$, then (4) and (16) yield

$$T(e_4, e_4) = n(m\,\gamma) = T^{44} = \text{energy density in } \{e_a\} \ . \tag{17}$$

In general we define an *energy-momentum 4-tensor* on $\mathcal{M}$ to be a symmetric, rank to 4-tensor T which satisfies

$$T(V,V) \geq 0 \tag{18}$$

for every timelike vector V ((18) is called the *weak energy condition*); for every *unit* timelike vector V, $T(V,V)$ is called the *energy density* measured in any admissible frame with $V = e_4$.

Exercise 2.3.5. Show that if T satisfies the weak energy condition, then (18) is also satisfied for any null vector V.

Exercise 2.3.6. An energy-momentum 4-tensor T is said to satisfy the *strong energy condition* if

$$T(V,V) \geq \frac{1}{2}(\text{trace } T)\,(V \cdot V) \ ,$$

for every timelike vector V. Show that this is true of the 4-tensor defined by (11). *Hint.* The *trace* of T is defined by trace $T = T^a_a = \eta_{ab}T^{ab}$ in each admissible coordinatization of $\mathcal{M}$. Show that $T(V,V) - \frac{1}{2}$ (trace T)$(V \cdot V) = n_0 m[(U \cdot V)^2 - (U \cdot U)(V \cdot V)]$ and then use the Reversed Schwartz Inequality (Theorem 1.5.4).

Exercise 2.3.7. Show that if T satisfies the strong energy condition, then (19) is also satisfied by any null vector V.

2.4. Electromagnetic Fields

We begin with a rather formidable sequence of definitions and then attempt some motivation. A 4-tensor F of rank 2 is said to be *skew-symmetric* if $F^{ba} = -F^{ab}$ for all $a,b, = 1,2,3,4$ (and so, in particular, $F^{aa} = 0$).

Exercise 2.4.1. Show that this definition is independent of the admissible basis relative to which the components of F are calculated.

Now let R be a region in $\mathcal{M}$. A (contravariant) *4-tensor field of rank 2 on R* is a function which assigns to every x in R a (contravariant) 4-tensor $F = F(x) = [F^{ab}(x^1,x^2,x^3,x^4)]_{a,b=1,2,3,4}$ of rank 2; the field is said to be *smooth* if each $F^{ab}(x^1,x^2,x^3,x^4)$ is C^∞ on R (i.e., has continuous partial derivatives of all orders and types with respect to x^1,x^2,x^3 and x^4 on R).

Exercise 2.4.2. Show that this definition of smoothness is independent of admissible coordinates.

For each $a,b,c = 1,2,3,4$, we let $F^{ab}{}_{,c} = \dfrac{\partial}{\partial x^c} F^{ab}$. It follows from the chain rule that, in another admissible coordinate system $(\bar{x}^a)$,

$$\bar{F}^{ab}{}_{,c} = \Lambda^a_\alpha \, \Lambda^b_\beta \, \bar{\Lambda}^\gamma_c \, F^{\alpha\beta}{}_{,\gamma} \, . \tag{20}$$

Remark. The transformation law (20) is characteristic of a "mixed 4-tensor of contravariant rank 2 and covariant rank 1". Such a 4-tensor can be identified in the obvious way with a trilinear real-valued function on 4-vectors ($T(U,V,W) = T^{ab}_c U_a V_b W^c$). Indeed, with the examples considered thus far the reader should have no difficulty formulating a general definition of a "4-tensor of contravariant rank r and covariant rank s" for any non-negative integers r and s and identifying such an object with an $(r+s)$ - linear real-valued function on 4-vectors, e.g., a 4-tensor of contravariant rank 1 and covariant rank 3 has components T^a_{bcd} in each admissible frame which transform according to $\bar{T}^a_{bcd} = \Lambda^a_\alpha \bar{\Lambda}^\beta_b \bar{\Lambda}^\gamma_c \bar{\Lambda}^\delta_d T^\alpha_{\beta\gamma\delta}$ and operates on four 4-vectors V,W,X and Y to give the real number

$$T(V,W,X,Y) = T^a_{bcd} V_a W^b X^c Y^d \, .$$

Now, an *electromagnetic field* on R is a smooth, skew-symmetric, contravariant 4-tensor field F of rank 2 on R whose components F^{ab} relative to any admissible basis for $\mathcal{M}$ satisfy *Maxwell's equations*

$$F^{ab}_{,a} = 0, \quad b = 1,2,3,4, \tag{21}$$

$$F^{ab}_{,c} + F^{bc}_{,a} + F^{ca}_{,b} = 0, \quad a,b = 1,2,3,4 \tag{22}$$

on R.

As to the motivation we ask the reader to recall that, in classical physics, an electromagnetic field was described in every admissible frame by two 3-vector fields

$\vec{E} = E_1 e_1 + E_2 e_2 + E_3 e_3$ (the "electric field vector") and $\vec{B} = B_1 e_1 + B_2 e_2 + B_3 e_3$ (the "magnetic field vector"). The components E_i and B_i are measured by well-defined and agreed upon experimental procedures and are found to satisfy the differential equations

$$\operatorname{div} \vec{E} = \frac{\partial E_1}{\partial x^1} + \frac{\partial E_2}{\partial x^2} + \frac{\partial E_3}{\partial x^3} = 0 \tag{23}$$

$$\operatorname{curl} \vec{B} - \frac{\partial \vec{E}}{\partial x^4} = \vec{0} \tag{24}$$

$$\operatorname{div} \vec{B} = \frac{\partial B_1}{\partial x^1} + \frac{\partial B_2}{\partial x^2} + \frac{\partial B_3}{\partial x^3} = 0 \tag{25}$$

$$\operatorname{curl} \vec{E} + \frac{\partial \vec{B}}{\partial x^4} = \vec{0} \tag{26}$$

in any *charge-free* region. Now, while one could seek a natural relativistic generalization of $\vec{E}$ and $\vec{B}$ separately, such an approach is clearly not recommended by the facts since how much of an "electromagnetic field" is electric and how much magnetic depends entirely upon the particular admissible frame in which the measurements are made. For instance, a single point charge at rest in frame $\mathcal{S}$ will give rise to a purely electrostatic field in $\mathcal{S}$, but, when viewed from another frame $\bar{\mathcal{S}}$ will be judged "moving" and consequently will generate a nonzero magnetic as well as an electric field (this is the familiar phenomenon of "electromagnetic induction"). A consistent relativistic view must therefore regard electric and magnetic fields as different manifestations of the same basic phenomenon. One therefore seeks a single "object" to act as the relativistic model of an electromagnetic field. The choice of a skew-symmetric 4-tensor of rank 2 is suggested by the fact that such a 4-tensor has precisely six independent components (just enough to accomodate E_1, E_2, E_3, B_1, B_2 and B_3). One possible arrangement of these six components in a 4×4 skew-symmetric matrix is

$$[F^{ab}] = \begin{bmatrix} 0 & B_3 & -B_2 & -E_1 \\ -B_3 & 0 & B_1 & -E_2 \\ B_2 & -B_1 & 0 & -E_3 \\ E_1 & E_2 & E_3 & 0 \end{bmatrix} \tag{27}$$

Reversing this point of view we formulate the following definitions. Given an electromagnetic field F on $R \subseteq \mathcal{M}$ we define, in each admissible frame, the real-valued functions E_i, B_i, $i = 1,2,3$, on R by

$$\begin{array}{lll} E_1 = -F^{14} & E_2 = -F^{24} & E_3 = -F^{34} \\ B_1 = F^{23} & B_2 = F^{31} & B_3 = F^{12} \end{array} \quad . \tag{28}$$

The *electric* and *magnetic 3-vector fields* associated with F in this frame are then defined by $\vec{E} = E_1 e_1 + E_2 e_2 + E_3 e_3$ and $\vec{B} = B_1 e_1 + B_2 e_2 + B_3 e_3$ respectively. With this notation, Maxwell's equations (21) and (22) reduce to (23) - (26).

Exercise 2.4.3. As sample verifications show that, when $b = 3$, (21) is just the third component of (24) and that (22) reduces to (25) when $a = 1$, $b = 2$, and $c = 3$. Also observe that, whenever two indices in (22) are equal, the equation is satisfied simply by virtue of the skew-symmetry of $[F^{ab}]$.

Remark. It is all very well to arrange the six components E_i and B_i in a skew-symmetric matrix such as (27), but it is not within our power to insist that these be the components of a 4-tensor which "represents" the electromagnetic field. In each admissible frame the electric and magnetic field components are *measured* so that whether or not the components in different frames are related by the transformation law (12) can only be determined by experiment. One can make such an experimental determination directly (based, for example, on (30) below) or one can assume the 4-tensor character of F^{ab} and judge the validity of this assumption by the success of the resulting theory.

It will be convenient to have the transformation law $\bar{F}^{ab} = \Lambda^a_\alpha \Lambda^b_\beta F^{\alpha\beta}$ written out explicitly in terms of the E_i and B_i, at least for the special Lorentz transformation (58) of Chapter 1. We ask the reader to perform the manual labor.

Exercise 2.4.4. Show that if Λ is the special Lorentz transformation (58) of Chapter 1, then, in terms of the notation established in (28), the transformation law (12) becomes

$$\begin{array}{lll} \bar{E}_1 = E_1 & \bar{E}_2 = \gamma(E_2 - \beta_r B_3) & \bar{E}_3 = \gamma(E_3 + \beta_r B_2) \\ \bar{B}_1 = B_1 & \bar{B}_2 = \gamma(B_2 + \beta_r E_3) & \bar{B}_3 = \gamma(B_3 - \beta_r E_2) \end{array} \tag{29}$$

The most important feature of (29) is the "mixing" of the field components. For example, when $\mathcal{S}$ measures a purely electric field ($B_1 = B_2 = B_3 = 0$), $\bar{\mathcal{S}}$ measures the components

$$\begin{array}{lll} \bar{E}_1 = E_1 & \bar{E}_2 = \gamma E_2 & \bar{E}_3 = \gamma E_3 \\ \bar{B}_1 = 0 & \bar{B}_2 = \gamma\beta_r E_3 & \bar{B}_3 = -\gamma\beta_r E_2 \end{array} \tag{30}$$

so that an observer in $\bar{\mathcal{S}}$ will, in general, experience a nonzero magnetic field. Similarly, if $E_1 = E_2 = E_3 = 0$, then

$$\begin{array}{lll} \bar{E}_1 = 0 & \bar{E}_2 = -\gamma\beta_r B_3 & \bar{E}_3 = \gamma\beta_r B_2 \\ \bar{B}_1 = B_1 & \bar{B}_2 = \gamma B_2 & \bar{B}_3 = \gamma B_3 \end{array} \tag{31}$$

Exercise 2.4.5. Show that $\frac{1}{2}F_{ab}F^{ab} = \vec{B} \cdot \vec{B} - \vec{E} \cdot \vec{E}$ in any admissible frame of reference and deduce that $\vec{B}^2 - \vec{E}^2$ is invariant under arbitrary Lorentz transformations. In particular, if $|\vec{E}| = |\vec{B}|$ in one frame, then the same must be true in any other frame. Here $F_{ab} = \eta_{\alpha a}\eta_{\beta b}F^{\alpha\beta}$ is the "purely covariant form" of F^{ab}.

We turn now to a description of the energy-momentum content of an electromagnetic field. Even in the classical theory of Maxwell an electromagnetic field was regarded as containing energy. By virtue of the mass-energy equivalence in relativity and the obvious intuitive analogy between an electromagnetic field and a "continuous distribution of mass-energy" one would expect this feature of the electromagnetic field to be modelled in relativity by an associated energy-momentum 4-tensor. The definition of the components of this 4-tensor in each admissible coordinatization of $\mathcal{M}$ will, of course, be based on the prescription laid down in (14). Moreover, since special relativity makes no changes in the mathematical details of electromagnetic theory in each admissible frame of reference (form invariance of Maxwell's equations under Lorentz transformations), all of the appropriate quantities ("energy density", "energy density flux", "stress", etc.) should be the same as the classical theory. In terms of the 3-vectors $\vec{E}$ and $\vec{B}$ these are

Energy Density

$$\frac{1}{8\pi}(\vec{E}^2 + \vec{B}^2)$$

Energy Density Flux (Poynting 3-Vector)

$$\frac{1}{4\pi}(\vec{E} \times \vec{B})$$

Maxwell's Stress 3-Tensor

$$\frac{1}{4\pi}\left(-(E_iE_j + B_iB_j) + \frac{1}{2}\delta^{ij}(\vec{E}^2 + \vec{B}^2)\right)$$

Thus, we need only define, for each electromagnetic field F^{ab}, an associated symmetric 4-tensor field T^{ab} whose components in each admissible coordinatization of $\mathcal{M}$ are given by (with the notation (28))

$$T^{44} = \frac{1}{8\pi}\,(\vec{E}^2 + \vec{B}^2)$$

$$T^{i4} = T^{4i} = \frac{1}{4\pi}\,(\vec{E} \times \vec{B}) \bullet e_i, \qquad i = 1,2,3$$

$$T^{ij} = T^{ji} = \frac{1}{4\pi}\,(-(E_iE_j + B_iB_j) + \frac{1}{2}\,\delta^{ij}(\vec{E}^2 + \vec{B}^2)), \quad i,j = 1,2,3.$$

We will show that this is accomplished by the following definition: Let F be an electromagnetic field on a region R of $\mathcal{M}$. Then the *energy-momentum 4-tensor* T *of* F has components in each admissible coordinatization of $\mathcal{M}$ defined by

$$T^{ab} = \frac{1}{4\pi}\,(\eta_{cd}F^{ad}F^{bc} - \frac{1}{2}\eta^{ab}\,(\vec{B}^2 - \vec{E}^2))\ . \tag{32}$$

Observe that T^{ab} is indeed a 4-tensor of the indicated type. In order to justify the definition we compute the components of T^{ab} in terms of the notation established in (28).

$$4\pi T^{44} = \eta_{cd}F^{4d}F^{4c} + \frac{1}{2}(\vec{B}^2 - \vec{E}^2)$$

$$= (F^{41}F^{41} + F^{42}F^{42} + F^{43}F^{43} - F^{44}F^{44}) + \frac{1}{2}\vec{B}^2 - \frac{1}{2}\vec{E}^2$$

$$= (E_1^2 + E_2^2 + E_3^2 - 0) + \frac{1}{2}\vec{B}^2 - \frac{1}{2}\vec{E}^2 = \vec{E}^2 + \frac{1}{2}\vec{B}^2 - \frac{1}{2}\vec{E}^2$$

$$= \frac{1}{2}\,(\vec{E}^2 + \vec{B}^2)\ .$$

so $T^{44} = \frac{1}{8\pi}\,(\vec{E}^2 + \vec{B}^2)$.

Next

$$4\pi T^{i4} = \eta_{cd}F^{id}F^{4c} - \frac{1}{2}\eta^{i4}(\vec{B}^2 - \vec{E}^2)$$

$$= \eta_{cd}F^{id}F^{4c} = F^{i1}F^{41} + F^{i2}F^{42} + F^{i3}F^{43} - F^{i4}F^{44}$$

$$= F^{i1}E_1 + F^{i2}E_2 + F^{i3}E_3\ .$$

For $i = 1$ this becomes

$$4\pi T^{14} = F^{11}E_1 + F^{12}E_2 + F^{13}E_3 = F^{12}E_2 + F^{13}E_3$$
$$= B_3E_2 - B_2E_3 = E_2B_3 - E_3B_2 = (\vec{E} \times \vec{B}) \cdot e_1$$

and, in general,

$$T^{i4} = T^{4i} = \frac{1}{4\pi}(\vec{E} \times \vec{B}) \cdot e_i, \quad i = 1,2,3.$$

Exercise 2.4.6. Show that, for all $i, j = 1,2,3$,

$$T^{ij} = T^{ji} = \frac{1}{4\pi}\left(-(E_iE_j + B_iB_j) + \frac{1}{2}\delta^{ij}(\vec{E}^2 + \vec{B}^2)\right).$$

Exercise 2.4.7. Show that an energy-momentum 4-tensor T for an electromagnetic field F has each of the following properties:

(a) trace $T = T^a_{\ a} = \eta_{ab}T^{ab} = 0$.

(b) T satisfies the strong energy condition.

(c) T satisfies the *conservation equations* $T^{ab}_{\ \ ,a} = 0$, $b = 1,2,3,4$.

Hint. Part (c) will require Maxwell's equations.

2.5. Charged Particles

Having introduced the appropriate mathematical device for modelling an electromagnetic field in $\mathcal{M}$ we now seek to model the physical agencies which produce such fields, i.e., electric charges. The definition is simple enough: A *charged particle* in $\mathcal{M}$ is a triple (α, m, e), where (α, m), $\alpha : I \to \mathcal{M}$, is a material particle and e is a nonzero real number called the *charge* of the particle. A *free charged particle* is defined in the same way, but with (α, m) a free material particle in $\mathcal{M}$. Charges do two things of interest to us. By their very presence they create electromagnetic fields and they also respond to the fields created by other charges. We shall conclude our study of relativistic mechanics by describing two examples of fundamental importance.

Consider first a *free* charged particle (α, m, e). Physically, such a charge creates an electromagnetic field and we must define a 4-tensor field F^{ab} on some region of $\mathcal{M}$ which models this field. Naturally, the precise form of this 4-tensor field must be postulated on the basis of experimental evidence. Our procedure will be as follows: $W = \alpha(I)$ is a timelike straight line in $\mathcal{M}$. For convenience we assume (without loss of generality) that W is a time axis. We may then select an admissible basis $\{e_a\}$ for $\mathcal{M}$ with $[e_4] = W$. Physically we view the charge as at rest in the corresponding reference frame $\mathcal{S}$. In order to specify the form of the electromagnetic field 4-tensor associated with (α, m, e) we postulate the form of the components in the rest frame $\mathcal{S}$ of the particle and decree that its components in any other admissible frame are to be computed from the 4-tensor transformation law (12). Explicitly, our definition is as follows: The electromagnetic field determined by the free charged particle (α, m, e), where $W = \alpha(I)$ is assumed a time axis, is the unique 4-tensor field on $\mathcal{M} - W$ whose contravariant components relative to a basis $\{e_a\}$ for which $[e_4] = W$ are given by *Coulomb's Law*:

$$E_1 = e\left(\frac{x^1}{r^3}\right) \qquad E_2 = e\left(\frac{x^2}{r^3}\right) \qquad E_3 = e\left(\frac{x^3}{r^3}\right) \tag{33}$$

$$B_1 = B_2 = B_3 = 0\ ,$$

where $r^2 = (x^1)^2 + (x^2)^2 + (x^3)^2$. Thus,

$$[F^{ab}] = e\begin{bmatrix} 0 & 0 & 0 & -\dfrac{x^1}{r^3} \\ 0 & 0 & 0 & -\dfrac{x^2}{r^3} \\ 0 & 0 & 0 & -\dfrac{x^3}{r^3} \\ \dfrac{x^1}{r^3} & \dfrac{x^2}{r^3} & \dfrac{x^3}{r^3} & 0 \end{bmatrix} \tag{34}$$

Remarks. Observe that the field F is singular on W. Also note that this definition does not depend on the choice of the basis $\{e_a\}$ with $[e_4] = W$ since any two bases with this property are related by a rotation in $\mathcal{L}$ (Lemma 1.3.4) and the definition (33) is invariant under rotations.

Exercise 2.5.1. Show that the 4-tensor field defined by (34) satisfies Maxwell's equations (21) and (22).

The definition (34), together with the 4-tensor character of the electromagnetic field and Maxwell's equations, lie at the very foundations of electromagnetic theory and can be used to solve a great variety of problems. Let us, for example, describe the electromagnetic field of a point charge e moving with uniform velocity β_r relative to some admissible frame $\mathfrak{I}$ at the instant the charge passes through the origin of this frame's spatial coordinate system. We may clearly assume without loss of generality that the charge moves along the $\bar{x}^1$-axis of $\mathfrak{I}$ in the negative direction. We can therefore accomplish our purpose by simply writing down the components of the field F for a free charge (α,m,e) in a reference frame related to the rest frame S of the charge by the special Lorentz transformation(58) of Chapter 1 and evaluating at $x^4 = \bar{x}^4 = 0$. Now, from (30) and (33) we obtain

$$\begin{array}{lll} \bar{E}_1 = e(\frac{x^1}{r^3}) & \bar{E}_2 = e\gamma(\frac{x^2}{r^3}) & \bar{E}_3 = e\gamma(\frac{x^3}{r^3}) \\ \bar{B}_1 = 0 & \bar{B}_2 = e\gamma\beta_r(\frac{x^3}{r^3}) & \bar{B}_3 = -e\gamma\beta_r(\frac{x^2}{r^3}) \end{array} \tag{35}$$

Now set $\bar{x}^4 = 0$ in (61) of Chapter 1 to obtain $x^1 = \gamma\bar{x}^1, x^2 = \bar{x}^2$ and $x^3 = \bar{x}^3$ so that $r^2 = \gamma^2 (\bar{x}^1)^2 + (\bar{x}^2)^2 + (\bar{x}^3)^2 =_{def} r'^2$. Thus, (35) becomes

$$\begin{array}{lll} \bar{E}_1 = e\gamma\frac{\bar{x}^1}{(r')^3} & \bar{E}_2 = e\gamma\frac{\bar{x}^2}{(r')^3} & \bar{E}_3 = e\gamma\frac{\bar{x}^3}{(r')^3} \\ \bar{B}_1 = 0 & \bar{B}_2 = e\gamma\beta_r\frac{\bar{x}^3}{(r')^3} & \bar{B}_3 = -e\gamma\beta_r\frac{\bar{x}^2}{(r')^3} \end{array} \tag{36}$$

We find then that the electric and magnetic 3-vectors in $\mathfrak{I}$ are

$$\vec{\bar{E}} = \frac{e\gamma}{(r')^3} (\bar{x}^1\bar{e}_1 + \bar{x}^2\bar{e}_2 + \bar{x}^3\bar{e}_3) = \frac{e\gamma}{(r')^3} \vec{\bar{x}} \tag{37}$$

$$\vec{\bar{B}} = \frac{e\gamma}{(r')^3} (0 \bullet \bar{e}_1 + \beta_r\bar{x}^3\bar{e}_2 - \beta_r\bar{x}^2\bar{e}_3) = \frac{e\gamma}{(r')^3} (\vec{\bar{u}} \times \vec{\bar{x}}) \ . \tag{38}$$

Observe that, in the non-relativistic limit ($\gamma \sim 1$), (37) and (38) reduce to

$$\vec{\bar{E}} = \frac{e}{\bar{r}^3} \vec{\bar{x}} \tag{39}$$

$$(\gamma \sim 1)$$

$$\vec{\bar{B}} = \frac{e}{\bar{r}^3} (\vec{\bar{u}} \times \vec{\bar{x}}) \tag{40}$$

The first equation shows that the electric field of a slowly moving charge is essentially that given by Coulomb's Law for a stationary charge, while (40) is the well-known *Biot-*

Savart Law.

Remark. It is interesting to observe that, according to (37), the electric field of a uniformly moving charge acts in a line with the point the charge occupies at the instant of measurement despite the fact that there is a nonzero time interval preceding that instant during which the behavior of the charge can no longer effect that measurement.

A charged particle in an electromagnetic field responds to the presence of that field by experiencing changes in 4-momentum. The precise quantitative nature of this response is expressed by an equation relating the proper time derivative of the particle's 4-momentum and the electromagnetic field 4-tensor (called the *equation of motion* of the charge). Such an equation of motion must, of course, be regarded as an addition to our basic Assumptions and is in no way a consequence of the principles of special relativity. That particular equation which is in closest accord with the experimental facts is the "Lorentz 4-force law" defined as follows: Let F be an electromagnetic field on a region R of $\mathcal{M}$ and let (α, m, e) be a charged particle with $W = \alpha(I) \subseteq R$. Then (α, m, e) is said to satisfy the *Lorentz 4-Force Law* with respect to F if

$$\frac{dP^a}{d\tau} = eF^a_b U^b, \quad a = 1,2,3,4, \tag{41}$$

where $U^b = \dfrac{dx^b}{d\tau}$ and $P^a = mU^a$ are the 4-velocity and 4-momentum of (α, m, e) and $F^a_b = \eta_{bc} F^{ac}$. As motivation for this definition we offer the following observations.

Exercise 2.5.2. Show that the spatial part ($a = 1,2,3$) of (41) can be written in each frame in terms of the particle's velocity 3-vector $\vec{v}$ and momentum 3-vector $\vec{P} = m\,\gamma\vec{v}$ and the electric and magnetic 3-vectors $\vec{E}$ and $\vec{B}$ as follows:

$$\frac{d\vec{P}}{dx^4} = e\,(\vec{E} + \vec{v} \times \vec{B}) \ . \tag{42}$$

Now, in reference frames relative to which the velocity of the particle is small (so that $\gamma \sim 1$ and the relative momentum 3-vector $\vec{P}$ is essentially the Newtonian momentum) equation (42) reduces to the classical equation of motion of a charge e under the influence of the "Lorentz force" $e\,(\vec{E} + \vec{v} \times \vec{B})$ due to an electromagnetic field. Observe also that when $a = 4$, (41) yields $\dfrac{dP^4}{d\tau} = e\,\gamma\,(E_1 v^1 + E_2 v^2 + E_3 v^3) = e\,\gamma \vec{E} \cdot \vec{v}$. Since $dP^4/d\tau = \gamma\, dP^4/dx^4$ and P^4 is the total relativistic energy E of (α, m, e) this may be written

$$\frac{dE}{dx^4} = e\,\vec{E} \bullet \vec{v}\,. \tag{43}$$

Note that the particle's energy change in each frame depends only on the electric components of F in that frame (the magnetic components do no work on the particle).

Remarks. Not every charged particle in an electromagnetic field satisfies the Lorentz 4-force law with respect to that field since other factors can influence the shape of the particle's worldline (e.g., it may be undergoing collisions). The equation of motion (41) is an appropriate model for charges whose motion is influenced *only* by the given electromagnetic field. Another proviso: (41) contains no term which reflects the contribution to the total electromagnetic field in R due to the particle (α, m, e) itself. In this context, (α, m, e) is regarded as a "test charge" whose own field must be negligible compared to F.

As an example we consider a charged particle (α, m, e) with $e > 0$ and an electromagnetic field F on $\mathcal{M}$ for which there exists an admissible basis $\{e_a\}$ relative to which $\vec{E} = \vec{0}$ and $\vec{B} = B_3 e_3$, with B_3 a positive constant (a constant magnetic field in the x^3-direction of some frame). Thus,

$$[F^{ab}] = \begin{bmatrix} 0 & B_3 & 0 & 0 \\ -B_3 & 0 & 0 & 0 \\ 0 & 0 & 0 & 0 \\ 0 & 0 & 0 & 0 \end{bmatrix}$$

We assume that (α, m, e) satisfies the Lorentz 4-force law with respect to F. Thus, the component equations $\dfrac{d^2x^a}{d\tau^2} = \dfrac{e}{m} F^a_b \dfrac{dx^b}{d\tau}$, $a = 1,2,3,4$, relative to the basis $\{e_a\}$ are (letting $\dfrac{e}{m} B_3 = \omega$)

$$\frac{d^2x^1}{d\tau^2} = \omega \frac{dx^2}{d\tau}$$

$$\frac{d^2x^2}{d\tau^2} = -\omega \frac{dx^1}{d\tau}$$

$$\frac{d^2x^3}{d\tau^2} = 0$$

$$\frac{d^2x^4}{d\tau^2} = 0$$

Exercise 2.5.3. By solving this system of differential equations show that there exist real numbers y^1, y^2, y^3, a, b, c and ϕ such that

$$(x^1, x^2, x^3, x^4) = (y^1 + a\sin(\omega\tau + \phi),\ y^2 + a\cos(\omega\tau + \phi),\ y^3 + b\tau,\ y^4 + c\tau). \quad (44)$$

Since (44) is the proper time parametrization of α we can compute the 4-velocity of the particle by differentiating component-wise with respect to τ:

$$(U^1, U^2, U^3, U^4) = (a\omega\cos(\omega\tau + \phi), -a\omega\sin(\omega\tau + \phi), b, c)\ . \quad (45)$$

Since $\eta_{ab}U^aU^b = -1$ we have $-1 = a^2\omega^2 + b^2 - c^2$ and therefore $c^2 = 1 + a^2\omega^2 + b^2$. Since $dx^4/d\tau = \gamma > 0$ we find that

$$(U^1, U^2, U^3, U^4) = (a\omega\cos(\omega\tau + \phi), -a\omega\sin(\omega\tau + \phi), b, (1 + a^2\omega^2 + b^2)^{\frac{1}{2}})\ . \quad (46)$$

We consider the case in which $b = 0$ so that (44) describes a worldline whose trajectory in S is a circle of radius $|a|$ in the x^1x^2-plane. In this case (46) becomes

$$(U^1, U^2, U^3, U^4) = (a\omega\cos(\omega\tau + \phi), -a\omega\sin(\omega\tau + \phi), 0, (1 + a^2\omega^2)^{\frac{1}{2}}) \quad (47)$$

$$= (\vec{v}, 1)\ .$$

Exercise 2.5.4. Compute $|\vec{v}|^2 = \beta^2$ from (47) and show that

$$\beta^2 = \frac{1}{\dfrac{m^2}{a^2e^2B_3^2} + 1}\ .$$

Now solve for $\dfrac{e}{m}$ to obtain the basic formula from which one computes the charge-to-mass ratio for particles in known uniform magnetic fields, e.g., in a bubble chamber:

$$\frac{e}{m} = \frac{1}{|a|B_3}\ \frac{\beta}{(1 - \beta^2)^{1/2}}\ .$$

PROBLEMS

2.A Aberration

A photon is observed in two admissible frames $\mathcal{S}$ and $\bar{\mathcal{S}}$, related by (59) of Chapter 1. The direction 3-vectors of the photon in $\mathcal{S}$ and $\bar{\mathcal{S}}$ are $\vec{d}$ and $\vec{\bar{d}}$, while the direction 3-vectors of $\bar{\mathcal{S}}$ relative to $\mathcal{S}$ and of $\mathcal{S}$ relative to $\bar{\mathcal{S}}$ are $\vec{d}$ and $\vec{\bar{d}}$ respectively. Define angles θ, $\bar{\theta}$ and θ' by $\cos\theta = \vec{d} \cdot d$, $\cos\bar{\theta} = \vec{\bar{d}} \cdot \vec{\bar{d}}$ and $\theta' = \pi - \bar{\theta}$.

1. Show that $\vec{\bar{d}} = -\gamma\,\beta_r\,(\beta_r e_4 + \vec{d})$.
2. Show that $\dfrac{\bar{\varepsilon}}{\varepsilon}\cos\bar{\theta} = \gamma\,(\beta_r - \cos\theta)$.
3. Use (8) to conclude that $\gamma\,(1 - \beta_r\cos\theta)\cos\bar{\theta} = \gamma(\beta_r - \cos\theta)$.
4. Derive the *relativistic aberration* formula:

$$\cos\theta' = \frac{\cos\theta - \beta_r}{1 - \beta_r\cos\theta}$$

2.B The Compton Effect

The physical situation we propose to model is the following: A photon collides with an electron and rebounds from it (generally with a different frequency), while the electron recoils from the collision. Thus, we consider a contact interaction $(\mathcal{A}, x, \mathcal{A}')$, where $\mathcal{A}$ consists of a photon with 4-momentum n and a material particle with proper mass m_e and 4-velocity U and $\mathcal{A}'$ consists of a photon with 4-momentum n' and a material particle with mass m_e and 4-velocity U'. We analyze the interaction in a frame of reference in which the material particle in $\mathcal{A}$ is at rest.

Figure 2.1

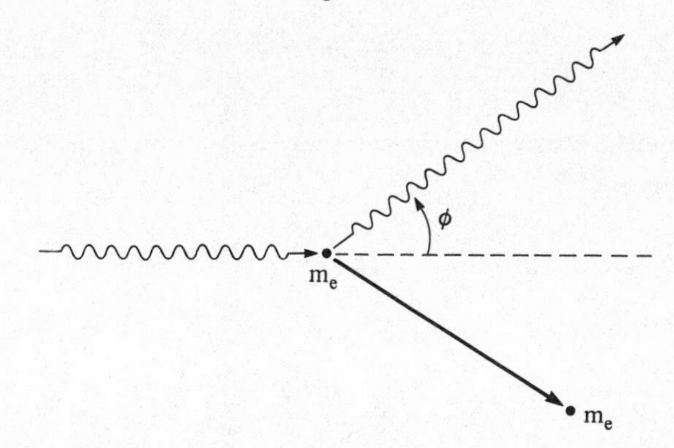

1. Show that, in this frame, the conservation of 4-momentum can be expressed in components as

$$m_e\gamma'(\mathrm{v}^i)' + h\nu'(l^i)' = h\nu l^i,\ i = 1,2,3 \tag{48}$$

$$m_e\gamma' + h\nu' = m_e + h\nu, \tag{49}$$

where $U = \gamma(\mathrm{v}^1,\mathrm{v}^2,\mathrm{v}^3,1)$, $U' = \gamma'((\mathrm{v}^1)',(\mathrm{v}^2)',(\mathrm{v}^3)',1)$, $\varepsilon = h\nu$, $\varepsilon' = h\nu'$ and l^i and $(l^i)'$ are the components of $\vec{d}$ and $\vec{d'}$ respectively.

2. Let $\xi = \nu'/\nu$, $k = h\nu/m_e$ and denote by ϕ the angle between the direction 3-vectors $\vec{d}$ and $\vec{d'}$ of the incident and rebounded photons, i.e., $\cos\phi = l^1(l^1)' + l^2(l^2)' + l^3(l^3)'$. Show that (48) and (49) can now be written

$$\gamma'(\mathrm{v}^i)' = kl^i - \xi k(l^i)',\ i = 1,2,3 \tag{50}$$

$$\gamma' - 1 = k(1-\xi)\ . \tag{51}$$

3. Let $(\beta')^2 = (\mathrm{v}^1)'(\mathrm{v}^1)' + (\mathrm{v}^2)'(\mathrm{v}^2)' + (\mathrm{v}^3)'(\mathrm{v}^3)'$, and show that $(\gamma')^2(\beta')^2 = k^2(1 - 2\xi\cos\phi + \xi^2) = (\gamma')^2 - 1$ so that (51) becomes

$$\gamma' + 1 = \frac{k^2(1 - 2\xi\cos\phi + \xi^2)}{\gamma' - 1} = \frac{k(1 - 2\xi\cos\phi + \xi^2)}{1-\xi}\ . \tag{52}$$

4. Subtract (51) from (52) to obtain

$$\xi = \frac{1}{1 + 2k\sin^2(\phi/2)}\ .$$

5. Show that the change in wavelength $\Delta\lambda = \dfrac{1}{\nu'} - \dfrac{1}{\nu}$ of the photon as a function of the angle ϕ through which it is deflected (in the frame under consideration) is

$$\Delta\lambda = \frac{2h}{m_e}\sin^2(\phi/2)$$

and observe that it does not depend on the wavelength λ of the incident photon. The maximum value

$$\Delta\lambda_{\max} = \frac{2h}{m_e}$$

of $\Delta\lambda$ occurs when $\phi = \pi$ and is a characteristic feature of the electron called its *Compton wavelength.*

2.C Inelastic Collision of Two Material Particles

We consider a contact interaction which is intended to model the following physical situation: Two material particles with masses m_1 and m_2 collide and coalesce to form a third material particle with mass m_3. Classically it is assumed that $m_3 = m_1 + m_2$ and on the basis of this assumption (and the conservation of Newtonian momentum) one finds that kinetic energy is lost during the collision. In Newtonian mechanics this lost kinetic energy disappears entirely from the mechanical picture in the sense that it is viewed as having taken the form of heat in the combined particle and therefore cannot be discussed further by the methods of mechanics. We shall see that this rather unsatisfactory feature of classical mechanics is avoided in relativistic mechanics by observing that the conservation of 4-momentum (which includes the conservation of energy) requires that the "hot" combined particle have a proper mass which is greater than the sum of the two masses from which it is formed, the difference $m_3 - (m_1 + m_2)$ being a measure of the energy required to bind the two particles together; this energy "acts like mass" in the combined particle. Let us then consider a contact interaction $(\mathcal{A}, x, \mathcal{A}')$, where $\mathcal{A}$ consists of two free material particles with proper masses m_1 and m_2 and 4-velocities U_1 and U_2 and $\mathcal{A}'$ consists of one free material particle with proper mass m_3 and 4-velocity U_3. Conservation of 4-momentum then requires that $m_3 U_3 = m_1 U_1 + m_2 U_2$ and the Reversed Triangle Inequality immediately gives $m_3 > m_1 + m_2$.

1. Show that, in any admissible frame,

$$m_3^2 = m_1^2 + m_2^2 + 2m_1 m_2 \gamma_1 \gamma_2 (1 - v_1^i v_2^i) \tag{53}$$

(summation over $i = 1,2,3$)

2. In any frame in which β_1 and β_2 are small one may approximate each γ_i by $1 + \frac{1}{2}\beta_i$. Then, retaining only second order terms, $\gamma_1 \gamma_2 \sim 1 + \frac{1}{2}\beta_1^2 + \frac{1}{2}\beta_2^2$. Derive the following approximate formula for m_3:

$$m_3^2 \sim (m_1 + m_2)^2 + m_1 m_2 (\beta_1^2 + \beta_2^2 - 2\gamma_1 \gamma_2 v_1^i v_2^i) \tag{54}$$

3. Now take $\gamma_1 \gamma_2 \sim 1$ in (54) and derive

$$m_3 \sim m_1 + m_2 + \frac{m_1 m_2}{m_1 + m_2 + m_3} \, |\vec{v}|^2 \tag{55}$$

where $\vec{v} = \vec{u}_1 - \vec{u}_2$. Assuming on the right-hand side that $m_3 \sim m_1 + m_2$ deduce that

$$m_3 \sim m_1 + m_2 + \frac{1}{2}\,\frac{m_1 m_2}{m_1 + m_2}\,|\vec{v}|^2 .$$

The term $\frac{1}{2}\,\frac{m_1 m_2}{m_1 + m_2}\,|\vec{v}|^2$ is the approximate gain in proper mass as a result of the collision and agrees with the Newtonian figure for the loss in kinetic energy.

2.D Uniqueness of the Energy-Momentum Tensor

Show that an energy-momentum 4-tensor is uniquely determined by its mass-energy density, i.e., that if S and T are two symmetric, second rank 4-tensors which satisfy $S^{44} = T^{44}$ in every admissible coordinate systems, then $S = T$.

Hint. Show that, for any future-directed, timelike vectors X and Y, $S^{ab}X_aY_b = T^{ab}X_aY_b$ and use the fact that $X + uY$ is future timelike for all u in some interval about $u = 0$.

2.E Timelike Eigenvectors

Show that an energy-momentum 4-tensor $[T^{ab}]$ has a timelike eigenvector if and only if there is an admissible observer who sees no net energy flux in any direction, i.e., $T^{41} = T^{42} = T^{43} = 0$.

2.F Pressure Free Perfect Fluids

In this problem we describe a "smoothed out" generalization of the particle stream which introduced section 2.3. We consider a region R in $\mathcal{M}$. A family of smooth curves with the property that every point of R is on one and only one curve in the family is called a *congruence* of curves in R. Let C denote a congruence of curves in R each of which is the worldline of a material particle. Let $\rho : R \to [0, \infty)$ be a smooth function R. If the energy-momentum 4-tensor for this system is defined at each point of R to be $T^{ab} = \rho U^a U^b$, where U is the 4-velocity vector field of the congruence, then the pair (C, ρ) is called a *pressure free perfect fluid* (or *dust*) with *energy density* ρ. Show that T^{ab} satisfies the strong energy condition at each point of R.

2.G Motion in a Constant $\vec{E}$-Field

Consider a free charged particle of mass m and charge e moving with constant speed β along the negative x^1-axis of some admissible frame. At $\tau = 0$ the particle is at this frame's origin ($x^1 = x^2 = x^3 = x^4 = 0$) and there encounters a constant $\vec{E}$-field in the x^3-direction ($\vec{E} = Ee_3$, $\vec{B} = \vec{0}$). Show tht the trajectory of the particle is given by

$$x^3 = \left(\frac{m\gamma}{eE}\right)(\cosh\left(\frac{eE}{m\beta\gamma}x^1\right) - 1),\ x^1 > 0\ ,$$

where $\gamma = (1 - \beta^2)^{-1/2}$.

2.H Hyperbolic Motion

In this problem we analyze the worldline of a material particle which experiences a constant "3-acceleration" relative to its "instantaneous rest frames". Consider then the worldline of a material particle parametrized by proper time τ with 4-velocity $U = U(\tau)$ and 4-acceleration $A = A(\tau)$. For each τ we let $\{e_a(\tau)\}$ be an admissible basis with $e_4(\tau)$ parallel to $U(\tau)$, i.e., an *instantaneous rest frame* for the particle at τ.

1. Show that, relative to its instantaneous rest frames, $U^4 = 1$, $A^4 = 0$, $U^i = dx^i/dx^4$ and $A^i = d^2x^i/(dx^4)^2$ for $i = 1,2,3$ and conclude that the squared magnitude of the acceleration 3-vector $\vec{A} = (d^2x^1/(dx^4)^2, d^2x^2/(dx^4)^2, d^2x^3/(dx^4)^2)$ is $A \cdot A$:

$$|\vec{A}|^2 = \sum_{i=1}^{3} (d^2x^i/(dx^4)^2)^2 = A \cdot A \ .$$

2. Assume that the magnitude of the particle's acceleration 3-vector in its rest frames is the constant g and that its motion is along the x^1-directions of these frames (i.e., $U^2 = U^3 = A^2 = A^3 = 0$). Write out the three conditions $U \cdot U = -1$, $U \cdot A = 0$ and $A \cdot A = g^2$ in coordinates and solve to obtain

$$A^1 = \frac{dU^1}{d\tau} = gU^4 \tag{56}$$

$$A^4 = \frac{dU^4}{d\tau} = gU^1 \ . \tag{57}$$

3. Show that U^1 satisfies the differential equation $\frac{d^2U^1}{d\tau^2} = g^2U^1$ and that the general solution to this equation is $U^1 = U^1(\tau) = a \sinh(g\tau) + b \cosh(g\tau)$. Assuming that the particle accelerates from rest at $\tau = 0$ ($U^1 = 0$ and $A^1 = g$ at $\tau = 0$), show that

$$U^1 = \sinh(g\tau) \ . \tag{58}$$

4. Deduce that

$$U^4 = \cosh(g\tau) \ . \tag{59}$$

5. Integrate (58) and (59) and assume that $x^2 = x^3 = x^4 = 0$ and $x_1 = 1/g$ at $\tau = 0$ to obtain $x^1 = \frac{1}{g}\cosh(g\tau)$ and $x^4 = \frac{1}{g}\sinh(g\tau)$.

6. Show that, on the particle's worldline, $(x^1)^2 - (x^2)^2 = 1/g^2$ so that this wordline is a hyperbola in the x^1x^4-plane.

CHAPTER 3

MORE GENERAL SPACETIMES: GRAVITY

3.1. Introduction

An electromagnetic field is a skew-symmetric, second rank 4-tensor field on $\mathcal{M}$ which satisfies Maxwell's equations (section 2.4). A charged particle responds to the presence of such a field by experiencing the changes in 4-momentum specified by the Lorentz 4-force law (section 2.5). This is how particle mechanics works. A physical agency which effects the shape of a particle's wordline is isolated and described mathematically and then equations of motion are postulated which quantify this effect. It would seem then that the next logical step in our program would be to carry out an analogous procedure for the gravitational field. In the early days of relativity theory many attempts to do so were made (by Einstein and others), some of great ingenuity, but they all came to nought. Whatever type of 4-tensor one selected to represent the gravitational field and however the corresponding field equations were chosen, the numbers simply did not come out right; theoretical predictions did not agree with the experimental facts (an account of some of these early attempts is available in Chapter 2 of [MTW]). Einstein soon turned away from the technician's task of formulating ever more refined variations in the hope of accomodating the observational data and sought instead an underlying physical reason for the failure of such apparently natural ideas. As always, the answer was there for anyone to see, but only Einstein saw it. An electromagnetic field is something "external" to the structure of spacetime, an additional field defined on and not influencing the mathematical structure of $\mathcal{M}$. Einstein realized that a gravitational field has a very special property which makes it unnatural to regard it as something external to the nature of the event world. Since Galileo it had been known that all objects with the same initial position and velocity respond to a given gravitational field in the same way (i.e., have identical worldlines) regardless of their material constitution (mass, charge, etc.). This is essentially what was verified at the Leaning Tower of Pisa and contrasts rather sharply with the behavior of electromagnetic fields. These worldlines (of particles with given initial conditions of motion) seem almost to be natural "grooves" in spacetime which anything will slide along if once placed there. But these "grooves" depend on the particular gravitational field being modeled and, in any case, $\mathcal{M}$ simply is not "grooved" (its structure does not distinguish any collection of curved worldlines). One suspects then that $\mathcal{M}$ itself is somehow lacking, that the appropriate mathematical structure for the event world may be more complex when gravitational effects are nonnegligible.

To see how the structure of $\mathcal{M}$ might be generalized to accomodate the presence of gravitational fields let us begin again as in section 1.2 with a structureless set M whose elements we call "events". One thing at least is clear: In regions that are distant from the source of any gravitational field, no accomodation is necessary and M must locally "look like" $\mathcal{M}$. But a great deal more is true. In his now famous "Elevator Experiment" Einstein observed that *any* event has about it a sufficiently small region of M which "looks like" $\mathcal{M}$. To see this we reason as follows: Imagine an elevator containing an observer and various other objects which is under the influence of some uniform external gravitational field. The cable snaps. The contents of the elevator are now in free fall. Since all of the objects inside respond to the gravitational field *in the same way* they will remain at relative rest throughout the fall. Indeed, if our observer lifts an apple from the floor and releases it in mid-air it will remain there. You have witnessed these facts for yourself. While it is unlikely you have ever had the misfortune of seeing a falling elevator you have seen astronauts at play inside their space capsules while in orbit (i.e., free fall) about the earth. The objects inside the elevator (capsule) seem then to constitute an archetypical inertial frame (they satisfy Newton's First Law). By establishing spatial and temporal coordinate systems in the usual way our observer thereby becomes an admissible observer, at least within the spatial and temporal constraints imposed by his circumstances. Now picture an arbitrary event. There are any number of vantage points from which the event can be observed. One is from a freely falling elevator in the immediate spatial and temporal vicinity of the event and from this vantage point the event receives *admissible* coordinates. There is then a *local admissible frame* near any event in M.

The operative word is "*local*". The "spatial and temporal constraints" to which we alluded in the preceding paragraph arise from the nonuniformity of any gravitational field in the real world. For example, in an elevator which falls freely in the earth's gravitational field, all of the objects inside are pulled toward the earth's *center* so that these objects do experience some slight relative motion (toward each other). Such motion, of course, goes unnoticed if the elevator falls neither too far nor too long. Indeed, by restricting our observer to a sufficiently small region in space and time these effects become negligible and the observer is indeed inertial. But then, what is "negligible" is in the eye of the beholder. The availability of more sensitive measuring devices will require further restrictions on the size of the spacetime region which "looks like" $\mathcal{M}$. Turn of the century mathematical terminology expressed this fact by saying that any point in M has about it an "infinitely small" neighborhood which is identical to $\mathcal{M}$, although in any "finite" region of M the deviations due to nonuniformity are (at least in principle) measurable. While this sort of terminology no doubt grates on the nerves of anyone trained in modern mathematics it should at least ring a bell. Indeed, it sounds very much like the terminology used in the 19th and early 20th centuries to describe smooth surfaces in $\mathbb{R}^3$ (any point on, for example, the sphere $x^2 + y^2 + z^2 = 1$ was said to have about it an "infinitely small" region which is identical to the plane $\mathbb{R}^2$). Spacetime is "locally like $\mathcal{M}$" in the same sense that the sphere is "locally like $\mathbb{R}^2$". Today we prefer to describe the situation in terms of local parameterization and the existence of tangent planes but the idea is the same. What seems to be emerging then as the appropriate mathematical structure for M is something analogous to a smooth surface, albeit a 4-dimensional one. As it happens there is in mathematics a notion (that of a "smooth manifold") which generalizes the definition of a smooth surface to higher dimensions. With each point in such a manifold is associated a flat "tangent space" and, just as for surfaces, the "curvature" of the manifold describes quantitatively the extent to which the manifold locally deviates from its tangent spaces, i.e., from flatness. In the particular manifolds of interest in relativity (called "Lorentz 4-manifolds" or "spacetimes") these deviations are taken to represent the effects of non-negligible gravitational fields. An object in free fall in such a field is represented by a curve that is "locally straight" since it would indeed appear straight in a nearby freely falling elevator (local inertial frame). On a surface in $\mathbb{R}^3$ the curves that are "locally straight" are its geodesics (e.g., the great circles on a sphere). In a spacetime manifold the analogous notion corresponds to the "grooves" to which we referred earlier.

3.2. Spacetimes

In this section we introduce the basic definitions from the theory of smooth manifolds. Here and throughout the chapter we present only that material that is indispensible to our study of relativity and opt always for the most elementary presentation of each concept.

We begin by reminding the reader of a few definitions from analysis. A subset U of $\mathbb{R}^n$, $n \geq 1$, is said to be *open* if each p in U is contained in some open ball $\{x \in \mathbb{R}^n \colon \| p - x \| < \varepsilon\}$, $\varepsilon > 0$, which is itself contained in U. An open set containing p is called a *neighborhood* of p. A subset K of $\mathbb{R}^n$ is *closed* if its complement $\mathbb{R}^n - K$ is open. A map $F \colon U \to \mathbb{R}^m$, $m \geq 1$, has *coordinate functions* $F_i \colon U \to \mathbb{R}$, $i = 1,2,\ldots,m$, defined by $F(p) = (F_1(p),...,F_m(p))$. The map F is *smooth* (or C^∞) if each F_i has continuous (partial) derivatives of all orders with respect to the variables $x^1,...,x^n$ in $U \subseteq \mathbb{R}^n$. If X is an arbitrary (not necessarily open) subset of $\mathbb{R}^n$, then a map $F: X \to \mathbb{R}^m$ is *smooth* if for each x in X there is an open subset U of $\mathbb{R}^n$ containing x and a C^∞ map $\hat{F}: U \to \mathbb{R}^m$ such that $\hat{F} \mid U \cap X = F \mid U \cap X$. A subset of X of the form $U \cap X$, where U is open in $\mathbb{R}^n$, is said to be *open in X*. A subset K of X is *closed in X* if $X - K$ is open in X. If $X \subseteq \mathbb{R}^n$ and $Y \subseteq \mathbb{R}^m$, then a map $F: X \to Y$ is a *diffeomorphism* if F is smooth, one-to-one and onto and the inverse map $F^{-1}: Y \to X$ is also smooth (if F and F^{-1} are only required to be continuous, then F is a *homeomorphism*). If a diffeomorphism of X onto Y exists we say that X and Y are *diffeomorphic* (*homeomorphic* in the continuous case). Finally, a subset X of $\mathbb{R}^n$ is *connected* if it cannot be written as $X = U_1 \cup U_2$, where U_1 and U_2 are open in X and disjoint.

Exercise 3.2.1. Suppose $X \subseteq \mathbb{R}^n$, $Z \subseteq X$ and $F: X \to \mathbb{R}^m$ is smooth. Show that the restriction $F \mid Z$ of F to Z is smooth as a map from Z into $\mathbb{R}^m$.

Exercise 3.2.2. Suppose $X \subseteq \mathbb{R}^n$, $Y \subseteq \mathbb{R}^m$, $Z \subseteq \mathbb{R}^l$ and $F: X \to Y$ and $G: Y \to Z$ are smooth. Show that the composition $G \circ F: X \to Z$ is smooth.

Exercise 3.2.3. Show that $F: \mathbb{R} \to \mathbb{R}$ defined by $F(x) = x^3$ is smooth one-to-one and onto, but is *not* a diffeomorphism. Is it a homeomorphism?

Exercise 3.2.4. Show that every open ball in $\mathbb{R}^n$ is diffeomorphic to $\mathbb{R}^n$. In particular, every open interval in $\mathbb{R}$ is diffeomorphic to $\mathbb{R}$. *Hint.* If $U_1(0) = \{p \in \mathbb{R}^n: \|p\| < 1\}$ is the open unit ball at the origin, then consider $F: U_1(0) \to \mathbb{R}^n$ defined by $F(p) = (1 - \|p\|^2)^{-1/2} p$.

Exercise 3.2.5. Let $U \subseteq \mathbb{R}^2$ be the open unit disc $\{(x^1,x^2) \in \mathbb{R}^2: (x^1)^2 + (x^2)^2 < 1\}$ and define $F: U \to \mathbb{R}^3$ by $F(x^1,x^2) = (x^1,x^2,(1-(x^1)^2-(x^2)^2)^{1/2})$. Show that F is a diffeomorphism onto its image and describe $F(U)$.

A connected subset M of some $\mathbb{R}^n$ is called a *k-dimensional smooth manifold* (or *smooth k-manifold*) if it is locally diffeomorphic to $\mathbb{R}^k$, i.e., if for each p in M there exists a connected open subset D of $\mathbb{R}^k$ and a map $\chi: D \to M$ such that $p \in \chi(D)$, $\chi(D)$ is open in M and χ is a diffeomorphism of D onto $\chi(D)$.

Exercise 3.2.6. Show that every point in a smooth k-manifold is contained in some open subset of M that is diffeomorphic to all of $\mathbb{R}^k$.

Such a local diffeomorphism χ is called a *coordinate patch* for M. The prototypes for this definition are the smooth 2-dimensional surfaces in $\mathbb{R}^3$. For example, each point on the 2-sphere $S^2 = \{(x^1,x^2,x^3) \in \mathbb{R}^3: (x^1)^2 + (x^2)^2 + (x^3)^2 = 1\}$ is contained in some open hemisphere which is the diffeomorphic image of a connected open subset of $\mathbb{R}^2$, e.g., the "upper hemisphere" $x^3 > 0$ is the image of the disc $(x^1)^2 + (x^2)^2 < 1$ under the diffeomorphism $\chi(x^1,x^2) = (x^1,x^2,(1-(x^1)^2-(x^2)^2)^{1/2})$ (see Exercise 3.2.5).

Exercise 3.2.7. Let D be a connected open subset of $\mathbb{R}^2$ and $f: D \to \mathbb{R}$ a C^∞ real-valued function on D. Show that the graph of f, i.e., $\{(x^1,x^2,f(x^1,x^2)): (x^1,x^2) \in D\}$, is a smooth 2-manifold. A 2-manifold such as this which is the image of a single coordinate patch is called a *simple surface* in $\mathbb{R}^3$.

The example of S^2 is easy to generalize: Let $f: \mathbb{R}^{k+1} \to \mathbb{R}$ be a C^∞ function. A point p in $\mathbb{R}^{k+1}$ is a *regular point* of f if at least one of the partial derivatives $\partial f/\partial x^i$, $i = 1,\ldots,k+1$, is nonzero at p. A real number r is a *regular value* of f if $f^{-1}(r)$ consists entirely of regular points.

Exercise 3.2.8. Show that if r is a regular value of $f: \mathbb{R}^{k+1} \to \mathbb{R}$, then $f^{-1}(r)$ is either empty or a k-dimensional smooth manifold in $\mathbb{R}^{k+1}$. *Hint.* You will need either the Inverse or Implicit Function Theorem.

A k-manifold in $\mathbb{R}^{k+1}$ of the form $f^{-1}(r)$, where r is a regular value of $f: \mathbb{R}^{k+1} \to \mathbb{R}$, is called a *level hypersurface* of f. In particular, the *k-sphere* $S^k = \{p \in \mathbb{R}^{k+1}: \|p\| = 1\}$ is a level hypersurface of the function $f(p) = \|p\|$.

Obviously, any open subset of $\mathbb{R}^k$ is a smooth k-manifold (take χ to be the identity map). In general, if M and N are smooth manifolds in $\mathbb{R}^n$ and $N \subseteq M$, then N is called a *submanifold* of M. In particular, M itself is a submanifold of its ambient Euclidean space $\mathbb{R}^n$.

One often produces new manifolds from old by forming products. Specifically, if M is a k-dimensional submanifold of $\mathbb{R}^n$ and N is an 1-dimensional submanifold of $\mathbb{R}^m$, then the Cartesian product $M \times N$ is a subset of $\mathbb{R}^n \times \mathbb{R}^m = \mathbb{R}^{n+m}$. We show now that it is, in fact, a $(k+1)$-dimensional submanifold of $\mathbb{R}^{n+m}$. Fix a point (p,q) in $M \times N$. Then there exists a connected open subset D_1 of $\mathbb{R}^k$ and a diffeomorphism χ_1 of D_1 onto an open subset.$\chi_1(D_1)$ of M containing p. Similarly, there exists a connected open subset D_2 of $\mathbb{R}^1$ and a diffeomorphism χ_2 of D_2 onto an open subset $\chi_2(D_2)$ of N containing q. We define a map $\chi_1 \times \chi_2$: $D_1 \times D_2 \to \mathbb{R}^n \times \mathbb{R}^m = \mathbb{R}^{n+m}$ by $\chi_1 \times \chi_2(x,y) = (\chi_1(x),\chi_2(y))$.

Exercise 3.2.9. Show that $D_1 \times D_2$ is a connected open subset of $\mathbb{R}^{k+1} = \mathbb{R}^k \times \mathbb{R}^1$ and that $\chi_1 \times \chi_2$ is a diffeomorphism of $D_1 \times D_2$ onto $\chi_1(D_1) \times \chi_2(D_2)$, which is open in $M \times N$.

We call the manifold $M \times N$ the *product* of M and N.

Exercise 3.2.10. Suppose $F: M \to M'$ and $G: N \to N'$ are smooth maps. Show that the *product map* $F \times G: M \times N \to M' \times N'$ defined by $F \times G\,(p,q) = (F\,(p), G\,(q))$ is smooth.

Exercise 3.2.11. Show that each of the *projection maps* π_M: $M \times N \to M$ and π_N: $M \times N \to N$ defined by $\pi_M(p,q) = p$ and $\pi_N(p,q) = q$ is smooth.

Some typical examples of product manifolds are the *torus* $S^1 \times S^1$ and the *cylinder* $S^1 \times \mathbb{R}$ in $\mathbb{R}^3$. Since all of our definitions and results extend immediately to products of any finite number of manifolds one can generalize and define, for example, higher dimensional tori $S^1 \times S^1 \times S^1$, etc.

Each point on a smooth surface in $\mathbb{R}^3$ has associated with it a 2-dimensional "tangent plane" consisting of the velocity vectors to all smooth curves on the surface through that point. The analogous construction on a smooth k-manifold M in $\mathbb{R}^n$ proceeds as follows: If $I \subseteq \mathbb{R}$ is an interval, then a continuous map α: $I \to M$ ($\subseteq \mathbb{R}^n$), $\alpha(t) = (x^1(t),...,x^n(t))$, is a *curve* in M. α is *smooth* if each coordinate function $x^i(t)$, $i = 1,...,n$, is C^∞ and if α's *velocity vector* $\alpha'(t) = (\frac{dx^1}{dt},...,\frac{dx^n}{dt})$ is nonzero for each t in I. Useful examples of smooth curves can be constructed from coordinate patches on M. Let χ: $D \to M$ be a coordinate patch on M, where D is a connected open subset of $\mathbb{R}^k$. Denoting the standard coordinates on $\mathbb{R}^k$ by $u^1,...,u^k$, we obtain the ith *coordinate curve* for χ from $\chi\,(u^1,...,u^k)$ by holding all u^j $(j \neq i)$ fixed; its velocity vector is denoted χ_i and is given by

$$\chi_i = \frac{\partial \chi}{\partial u^i} = \frac{\partial}{\partial u^i}\,(x^1(u^1,...,u^k),...,x^n(u^1,...,u^k))$$

$$= \left(\frac{\partial x^1}{\partial u^i},...,\frac{\partial x^n}{\partial u^i}\right).$$

The coordinate curves of χ cover $\chi\,(D)$ with a "coordinate system" so that each p in $\chi\,(D)$ is uniquely specified by k coordinates (the Cartesian coordinates of $\chi^{-1}\,(p)$).

Figure 3.1

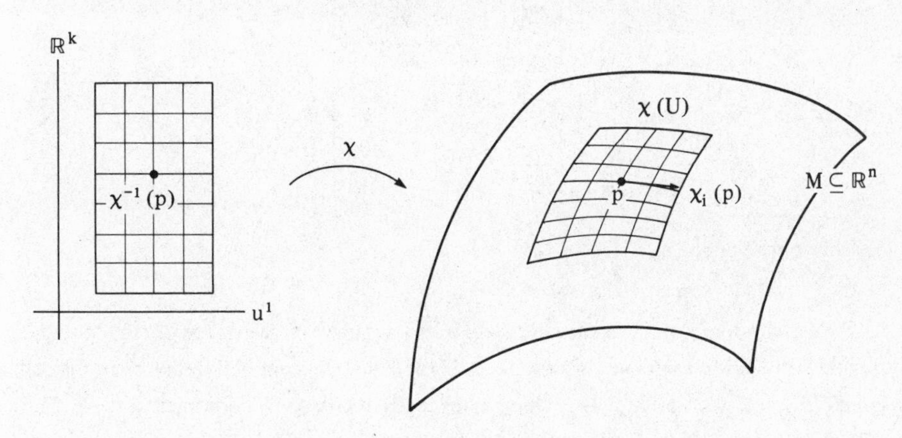

In particular, any smooth real-valued function defined (at least) on $\chi(D)$ can be regarded as a function of these k coordinates by composing with χ: $f(u^1,...,u^k) = f \circ \chi$

Exercise 3.2.12. Write down a coordinate patch for the 4-sphere S^4 and calculate its coordinate velocity vectors χ_i, $i = 1,2,3,4$.

Using the definition alone it can be quite difficult to determine whether or not a given smooth map of $\mathbb{R}^k$ into M is a coordinate patch. Fortunately, there is a remarkable theorem in advanced calculus to which we can appeal for assistance. In its usual form the Inverse Function Theorem asserts that if F is a smooth map between open subsets of $\mathbb{R}^k$ and if the Jacobian of F is nonsingular at some point p, then F carries some open neighborhood of p in the domain diffeomorphically onto some open neighborhood of $F(p)$ in the range (see, e.g., [Sp1]). We leave it to the reader to persuade himself that the version we require follows directly from this:

Theorem 3.2.1. Let $\chi: D \to M$ be smooth, where D is a connected open subset of $\mathbb{R}^k$ and $M \subseteq \mathbb{R}^n$ is a smooth k-manifold. Define $\chi_1,...,\chi_k$ at each point of D by $\chi_i = \partial\chi/\partial u^i = (\partial x^1/\partial u^i,...,\partial x^n/\partial u^i)$ and let p be a point of D. Then χ is a diffeomorphism of some open subset U of D containing p onto some open subset V of M containing $F(p)$ if and only if $\chi_1(p),...,\chi_k(p)$ are linearly independent, i.e., if and only if the Jacobian

$$\begin{bmatrix} \frac{\partial x^1}{\partial u^1} & \cdots & \frac{\partial x^1}{\partial u^k} \\ \cdot & & \cdot \\ \cdot & & \cdot \\ \cdot & & \cdot \\ \frac{\partial x^n}{\partial u^1} & \cdots & \frac{\partial x^n}{\partial u^k} \end{bmatrix}$$

of F has rank k at p.

Now consider a smooth curve $\alpha: I \to M$ and assume that the image of α is contained in $\chi\,(D)$ for some coordinate patch $\chi: D \to M$ (note that, by continuity, each t_0 in I is contained in some subinterval J of I which maps entirely into some coordinate patch). Then $\chi^{-1} \circ \alpha(t)$ is a smooth curve $t \to (u^1(t),...,u^k(t))$ in D so α can be written

$$\alpha(t) = \chi(u^1(t),...,u^k(t)) \ .$$

By the Chain Rule, $\alpha'(t) = \frac{d}{dt}\,\chi(u^1(t),...,u^k(t)) = \sum_{i=1}^{k} \frac{\partial \chi}{\partial u^i}\,\frac{du^i}{dt}$, i.e., (using the summation convention),

$$\alpha'(t) = (u^i(t))'\,\chi_i(\alpha(t)) = (u^i)'\,\chi_i \ . \tag{1}$$

At each point p in M the *tangent space* to M at p, denoted $T_p(M)$, is the set of all velocity vectors at p to smooth curves in M through p (together with the zero vector). The elements of $T_p(M)$ are called *tangent vectors* to M at p and each is a linear combination of the coordinate velocity vectors $\{\chi_i(p)\}_{i=1}^{k}$ for any coordinate patch χ with p in $\chi(D)$.

Exercise 3.2.13. Show that, conversely, any nontrivial linear combination of $\chi_1(p),...,\chi_k(p)$ is the velocity vector to some smooth curve in M through p.

Since χ is a diffeomorphism the χ_i, being columns of the Jacobian, are linearly independent. Consequently, any coordinate patch at p gives rise to a basis for $T_p(M)$ consisting of the coordinate velocity vectors at p. In particular, $T_p(M)$ is a k-dimensional vector space (or affine space if you prefer to picture $T_p(M)$ at p rather than at 0 in $\mathbb{R}^n$).

Since we have determined that the event world is "locally like $\mathcal{M}$" at each of its points we elect to model it by a smooth 4-manifold whose tangent spaces are all provided with the structure of Minkowski spacetime, i.e., a Lorentz inner product. A smooth assignment of an inner product to each tangent space of a manifold M is called a "metric" on M (not to be confused with the term used in point-set topology for a "distance function", although there are some connections; see Problem 4.C). More precisely, a *metric* (or *metric tensor*) g on the smooth k-manifold M is an assignment to each tangent space $T_p(M)$ of an inner product $g_p = <,>_p$ such that the *component functions*

$$g_{ij}(p) = g_p(\chi_i(p), \chi_j(p)) = <\chi_i(p), \chi_j(p)>_p$$

are C^∞ on $\chi(D)$ for each coordinate patch χ at p. If each inner product g_p is of index one, then the metric g is called a *Lorentz metric*; if each g_p has index zero (i.e., is positive definite), then g is a *Riemannian metric*. A *spacetime* is a smooth 4-manifold on which is defined a Lorentz metric.

Remark. We shall often abuse our notation by omitting references to "p". Thus, $g_p(v,w) = <v,w>_p = g(v,w) = <v,w>$ and on occasion we may use $v \bullet w$ for variety. For the sake of economy we shall also refer to any manifold on which is defined either a Lorentz or a Riemannian metric as a "manifold with metric".

The assumption that the metric g of a spacetime has index one means that each $T_p(M)$ has a basis $\{e_1, e_2, e_3, e_4\}$ such that $g(e_i, e_j) = \eta_{ij}$. On the other hand, relative to a coordinate basis $\{\chi_i(p)\}$ corresponding to some coordinate patch, $g(\chi_i, \chi_j) = g_{ij}$ so that if $\mathrm{v} = \mathrm{v}^i \chi_i$ and $w = w^j \chi_j$, then

$$g(\mathrm{v}, w) = g(\mathrm{v}^i \chi_i, w^j \chi_j) = g_{ij} \mathrm{v}^i w^j \ .$$

Examples of Riemannian manifolds are easy to write down by simply restricting the usual dot product of $\mathbb{R}^n$ to each $T_p(M)$.

Exercise 3.2.14. Let $D = \{(u^1, u^2) \in \mathbb{R}^2 : -\pi < u^1 < \pi,\ 0 < u^2 < \pi\}$ and suppose R is a positive constant. Define a map χ from D into $\mathbb{R}^3$ by $\chi(u^1, u^2) = (R \cos u^1 \sin u^2,\ R \sin u^1 \sin u^2,\ R \cos u^2)$.

(a) Show that the image of χ is a portion of the sphere of radius R about the origin in $\mathbb{R}^3$. What portion? *Hint.* u^1 and u^2 are spherical coordinates.

(b) Show that the coordinate velocity vectors χ_i, $i = 1,2$, are $\chi_1 = (-R \sin u^1 \sin u^2,\ R \cos u^1 \sin u^2,\ 0)$ and $\chi_2 = (R \cos u^1 \cos u^2,\ R \sin u^1 \cos u^2,\ -R \sin u^2)$.

(c) Show that χ is a coordinate patch for the sphere and describe the u^1- and u^2-coordinate curves.

(d) Define a Riemannian metric on $\chi(D)$ by restricting the usual dot product of $\mathbb{R}^3$ to each tangent space and show that

$$\begin{bmatrix} g_{11} & g_{12} \\ g_{21} & g_{22} \end{bmatrix} = \begin{bmatrix} R^2 \sin^2 u^2 & 0 \\ 0 & R^2 \end{bmatrix}.$$

Exercise 3.2.15. Let $D = \{(u^1, u^4) \in \mathbb{R}^2 : -\pi < u^1 < \pi, -\infty < u^4 < \infty\}$ (the reason for the peculiar numbering will become clear in Exercise 3.2.18). Define $\chi: D \to \mathbb{R}^3$ by $\chi(u^1, u^4) = (\cosh u^4 \cos u^1, \cosh u^4 \sin u^1, \sinh u^4)$.

(a) Show that the image of χ is a portion of the hyperboloid of one sheet $(x^1)^2 + (x^2)^2 - (x^3)^2 = 1$. What portion?

(b) Show that χ is a coordinate patch and describe the u^1- and u^4-coordinate curves.

(c) Define a Riemannian metric on $\chi(D)$ by restricting the $\mathbb{R}^3$ dot product to each tangent space and show that

$$\begin{bmatrix} g_{11} & g_{14} \\ g_{41} & g_{44} \end{bmatrix} = \begin{bmatrix} \cosh^2 u^4 & 0 \\ 0 & -1 \end{bmatrix}$$

Remark. We wish to make the reader aware of a notational device which we have chosen not to employ, but which is popular in the literature of our subject. In a surface such as the sphere a smooth curve $\alpha: I \to M$ with $\alpha(I) \subseteq \chi(D)$ has velocity vector $\alpha'(t) = (u^i(t))' \chi_i(\alpha(t))$. The squared "speed" of α is then $g(\alpha'(t), \alpha'(t)) = g_{ij}(u^i)'(u^j)' = g_{ij} \dfrac{du^i}{dt} \dfrac{du^j}{dt}$. One then defines the "arc length" $s = s(t)$ along α as usual by $(\dfrac{ds}{dt})^2 = g_{ij} \dfrac{du^i}{dt} \dfrac{du^j}{dt}$ and this is often written in "differential form" by suppressing all references to the parameter t:

$$ds^2 = g_{ij}\, du^i\, du^j .$$

This last expression is often referred to as "the metric" of M. For example, using the more familiar notation $u^1 = \theta$ and $u^2 = \varnothing$ one would regard

$$ds^2 = R^2 \sin^2 \varnothing\, d\theta^2 + R^2\, d\varnothing^2$$

as the metric of the sphere of radius R. We prefer to regard such expressions as simply a convenient way of displaying the matrix (g_{ij}) on a single line and of remembering how to compute arc lengths.

At this point we begin to develop a basic stock of examples of spacetimes to which we shall return repeatedly throughout the text. Along the way we will need to introduce some of the most fundamental notions of modern differential geometry. First though, the obvious definitions: If M is a spacetime, then a tangent vector v is *spacelike, timelike* or *null* according as $g(\mathrm{v},\mathrm{v})$ is >0, <0 or $=0$ respectively. The *null cone* at p in M is the subset $\{\mathrm{v} \in T_p(M): g(\mathrm{v},\mathrm{v}) = 0\}$ of $T_p(M)$.

The most obvious example is, of course, $\mathcal{M}$ itself. Specifically, we let $D = \mathbb{R}^4$ and take χ to be the identity map on $\mathbb{R}^4$, i.e., $\chi(u^1,u^2,u^3,u^4) = (u^1,u^2,u^3,u^4)$ $(x^i(u^1,u^2,u^3,u^4) = u^i,\ i = 1,2,3,4)$. Denoting $\chi(D)$ by $\mathcal{M}$ we have $\chi_1 = (1,0,0,0)$, $\chi_2 = (0,1,0,0)$, $\chi_3 = (0,0,1,0)$ and $\chi_4 = (0,0,0,1)$ so that, in effect, each $T_p(\mathcal{M})$ can be identified in the obvious way with $\mathbb{R}^4$. We define a metric g on $\mathcal{M}$ by specifying its component functions $g_{ij}(p)$ relative to $\{\chi_i(p)\}_{i=1}^4$. As expected we take

$$g_{ij}(p) = g(\chi_i(p), \chi_j(p)) = \eta_{ij} .$$

Of course, the $g_{ij}(p)$ are, in this case, constant, i.e., independent of p.

Just as Cartesian coordinates are not always the most convenient choice in $\mathbb{R}^3$ so this standard coordinate patch for $\mathcal{M}$ is often replaced by one more naturally adapted to the problem at hand. Before describing one such we discuss, in general, the relationship between two overlapping coordinate systems on an arbitrary manifold M. Thus, we consier two coordinate patches $\chi: D \to M$ and $\bar{\chi}: \bar{D} \to M$ and assume that $\chi(D) \cap \bar{\chi}(\bar{D}) \neq \emptyset$.

Figure 3.2

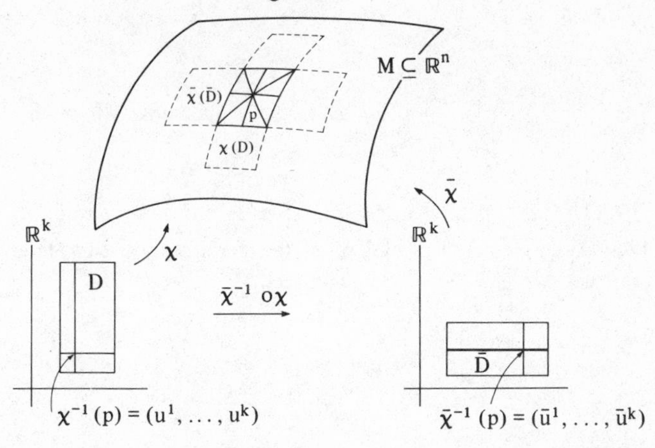

A given p in $\chi(D) \cap \bar{\chi}(\bar{D})$ now has two sets of coordinates $\chi^{-1}(p) = (u^1, ..., u^k)$ and $\bar{\chi}^{-1}(p) = (\bar{u}^1, ..., \bar{u}^k)$. Observe that the composition $\bar{\chi}^{-1} \circ \chi$ is smooth on $\chi^{-1}(\chi(D) \cap \bar{\chi}(\bar{D}))$, i.e., the $\bar{u}^i$ are C^∞ functions of the u^j:

$$\bar{u}^i = \bar{u}^i(u^1, ..., u^k), \quad i = 1, ..., k \ .$$

Moreover,

$$\chi(u^1, ..., u^k) = \bar{\chi}(\bar{u}^1(u^1, ..., u^k), ..., \bar{u}^k(u^1, ..., u^k)) \ .$$

Exercise 3.2.16. Show that at each point of $\chi(D) \cap \bar{\chi}(\bar{D})$,

$$\chi_i = \frac{\partial \bar{u}^j}{\partial u^i} \bar{\chi}_j, \quad i = 1, ..., k \tag{2}$$

and that, for each tangent vector v at a point in the intersection, if $\mathrm{v} = \mathrm{v}^i \chi_i$ and $\mathrm{v} = \bar{\mathrm{v}}^j \bar{\chi}_j$, then

$$\bar{\mathrm{v}}^j = \frac{\partial \bar{u}^j}{\partial u^i} \mathrm{v}^i, \quad j = 1, ..., k \ . \tag{3}$$

Finally, show that the metric component functions g_{ij} and $\bar{g}_{ij}$ in the two coordinate systems are related by

$$\bar{g}_{ij} = \frac{\partial u^1}{\partial \bar{u}^i} \frac{\partial u^m}{\partial \bar{u}^j} g_{1m}, \quad i, j = 1, ..., k \ . \tag{4}$$

Now let us define a new coordinate patch for (a part of) $\mathcal{M}$ by employing spherical spatial coordinates $(\bar{u}^1 = \rho, \bar{u}^2 = \phi, \bar{u}^3 = \theta, \bar{u}^4 = u^4)$, i.e., we let $\bar{D} = \{(\bar{u}^1, \bar{u}^2, \bar{u}^3, \bar{u}^4) \in \mathbb{R}^4 : \bar{u}^1 > 0, 0 < \bar{u}^2 < \pi, -\pi < \bar{u}^3 < \pi, -\infty < \bar{u}^4 < \infty\}$ and define $\bar{\chi}: \bar{D} \to \mathcal{M}$ by

$$\bar{\chi}(\bar{u}^1, \bar{u}^2, \bar{u}^3, \bar{u}^4) = (\bar{u}^1 \cos \bar{u}^3 \sin \bar{u}^2, \bar{u}^1 \sin \bar{u}^3 \sin \bar{u}^2, \bar{u}^1 \cos \bar{u}^2, \bar{u}^4) \ .$$

Exercise 3.2.17. Show that $\bar{\chi}$ is a coordinate patch for a portion of $\mathcal{M}$. What portion? Show, in two ways, that on this portion of $\mathcal{M}$, the metric components are given by

$$\bar{g}_{11} = 1, \ \bar{g}_{22} = (\bar{u}^1)^2, \ \bar{g}_{33} = (\bar{u}^1)^2 \sin^2 \bar{u}^2 ,$$

$$\bar{g}_{44} = -1 \ \ \bar{g}_{ij} = 0 \ \text{if} \ i \neq j \ .$$

Our second example is called *deSitter spacetime*, is denoted $\mathcal{D}$ and will be identified as a manifold with a level hypersurface in $\mathbb{R}^5$. Indeed, the metric on $\mathcal{D}$ will be the restriction to $\mathcal{D}$ of the "Minkowski inner product" on $\mathbb{R}^5$. Specifically, let us denote by $\mathbb{R}_1^5$ the linear space $\mathbb{R}^5$ with the inner product g_1 defined by

$$g_1(x,y) = x^1y^1 + x^2y^2 + x^3y^3 + x^4y^4 - x^5y^5$$

where $x = (x^1,x^2,x^3,x^4,x^5)$ and $y = (y^1,y^2,y^3,y^4,y^5)$ (see Exercise 1.1.1). The associated quadratic form $Q_1 : \mathbb{R}^5 \to \mathbb{R}$ is defined, as usual, by

$$Q_1(x) = (x^1)^2 + (x^2)^2 + (x^3)^2 + (x^4)^2 - (x^5)^2$$

and is obviously smooth on all of $\mathbb{R}^5$. Since $\partial Q_1/\partial x^i = 2x^i$ for $i = 1,2,3,4$ and $\partial Q_1/\partial x^5 = -2x^5$ it is clear that $r = 1$ is a regular value of Q_1 so that $Q_1^{-1}(1) = \{x \in \mathbb{R}^5 : (x^1)^2 + (x^2)^2 + (x^3)^2 + (x^4)^2 - (x^5)^2 = 1\}$ is a 4-dimensional smooth submanifold of $\mathbb{R}^5$ (Exercise 3.2.8). We set $\mathcal{D} = Q_1^{-1}(1)$ and picture $\mathcal{D}$ as a "hyperboloid of one sheet" in $\mathbb{R}^5$ by suppressing x^3 and x^4 as in Figure 3.3.

Figure 3.3

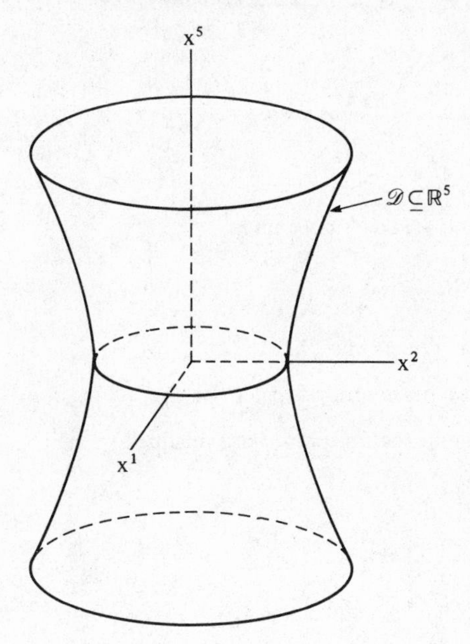

Mentally re-inserting the missing dimensions it becomes clear that the cross-sectional "circles" in Figure 3.3 are actually 3-spheres S^3 so that $\mathcal{D}$, as a manifold, is $S^3 \times \mathbb{R}$.

Theorem 3.2.2. deSitter spacetime $\mathcal{D}$ is diffeomorphic to $S^3 \times \mathbb{R}$.

Proof: $S^3 \times \mathbb{R} = \{(y^1,y^2,y^3,y^4,t) : (y^1)^2 + (y^2)^2 + (y^3)^2 + (y^4)^2 = 1, -\infty < t < \infty\}$. Define a map $F : S^3 \times \mathbb{R} \to \mathbb{R}^5$ by $F(y^1,y^2,y^3,y^4,t) = ((1+t^2)^{1/2}\, y^1,$ $(1+t^2)^{1/2}\, y^2$, $(1+t^2)^{1/2}\, y^3$, $(1+t^2)^{1/2}\, y^4$, $t)$. Then F is clearly smooth and one-to-one and, moreover, $Q_1(F(y^1,y^2,y^3,y^4,t)) = 1$ so F maps into $\mathcal{D}$. Since the map $G : \mathcal{D} \to S^3 \times \mathbb{R}$ defined by $G(x^1,x^2,x^3,x^4,x^5) =$ $= ((1+(x^5)^2)^{1/2}\, x^1, (1+(x^5)^2)^{1/2}\, x^2, (1+(x^5)^2)^{1/2}\, x^3, (1+(x^5)^2)^{1/2}\, x^4, x^5)$ is smooth on all $\mathcal{D}$ and is easily seen to be an inverse for F, F is a diffeomorphism. Q.E.D.

Now we shall define a Lorentz metric g on $\mathcal{D}$ by restricting the $\mathbb{R}^5_1$-inner product g_1 to each $T_p(\mathcal{D})$.

Exercise 3.2.18. Let $D = \{(u^1,u^2,u^3,u^4) \in \mathbb{R}^4 : -\pi < u^1 < \pi,\ 0 < u^2 < \pi,$ $-\pi < u^3 < \pi,\ -\infty < u^4 < \infty\}$ define $\chi : D \to \mathbb{R}^5$ by $\chi(u^1,u^2,u^3,u^4) = (x^1,x^2,x^3,x^4,x^5)$, where

$$x^1 = \cosh u^4 \cos u^1$$

$$x^2 = \cosh u^4 \sin u^1 \cos u^2$$

$$x^3 = \cosh u^4 \sin u^1 \sin u^2 \cos u^3$$

$$x^4 = \cosh u^4 \sin u^1 \sin u^2 \sin u^3$$

$$x^5 = \sinh u^4$$

(a) Show that χ maps into $\mathcal{D}$.

(b) Show tht χ is a coordinate patch for $\mathcal{D}$.

(c) Show that the metric components relative to χ are

$$g_{11} = \cosh^2 u^4$$

$$g_{22} = \cosh^2 u^4 \sin^2 u^1$$

$$g_{33} = \cosh^2 u^4 \sin^2 u^1 \sin^2 u^2$$

$$g_{44} = -1$$

$$g_{ij} = 0 \quad \text{if } i \neq j\ .$$

If u^1, u^2 and u^3 are confined to the closed intervals $[-\pi,\pi]$, $[0,\pi]$ and $[-\pi,\pi]$ respectively, then the map χ in Exercise 3.2.18 maps *onto* $\mathcal{D}$. Thus, by employing the same map χ, but with u^1, u^2 and u^3 restricted to sufficiently small open intervals about $-\pi, 0$ and π one obtains finitely many coordinate patches which cover $\mathcal{D}$. Our next example is even more manageable in that, like $\mathcal{M}$, it is the image of a single coordinate patch. The *Einstein-deSitter spacetime* $\mathcal{E}$ is the simplest of all the "cosmological models" to emerge from general relativity and is defined as follows: Let $D = \mathbb{R}^3 \times (0,\infty) \subseteq \mathbb{R}^4$, take χ to be the identity map on D and let $\mathcal{E} = \chi(D) = \mathbb{R}^3 \times (0,\infty)$. Then $\chi_1 = (1,0,0,0)$, $\chi_2 = (0,1,0,0)$, etc. at each point p in $\mathcal{E}$. We describe the Lorentz metric g on $\mathcal{E}$ by giving its components relative to the basis $\{\chi_i(p)\}$ at each $p = (u^1,u^2,u^3,u^4)$ in $\mathcal{E}$. Specifically, we set

$$g_{ij}(p) = g_{ij}(u^1,u^2,u^3,u^4) = \begin{cases} 0 & \text{if } i \neq j \\ (u^4)^{4/3} & \text{if } i = j = 1,2,3 \\ -1 & \text{if } i = j = 4 \end{cases}$$

Thus, for $\mathrm{v}, w \in T_p(\mathcal{E})$, $g_p(\mathrm{v},w) = g_p(\mathrm{v}^i\chi_i, w^j\chi_j) = (u^4)^{4/3}\,(\mathrm{v}^1w^1 + \mathrm{v}^2w^2 + \mathrm{v}^3w^3) - \mathrm{v}^4w^4$, where u^4 is the "height" of p. Note that this *is* a Lorentz metric since smoothness is obvious on $u^4 > 0$ and, for each $p \in \mathcal{E}$, one can define a new basis $\{e_i\}_{i=1}^4$ for $T_p(\mathcal{E})$ by $e_i = (u^4)^{-2/3}\,\chi_i$ for $i = 1,2,3$ and $e_4 = \chi_4$; then $g(e_i,e_j) = \eta_{ij}$. Next observe that for $p \in \mathcal{E}$, the null cone at p is $\{\, \mathrm{v} = \mathrm{v}^i\chi_i \in T_p(\mathcal{E})\colon (u^4)^{4/3}\,((\mathrm{v}^1)^2 + (\mathrm{v}^2)^2 + (\mathrm{v}^3)^2) = (\mathrm{v}^4)^2 \}$ which one might interpret geometrically as saying that the null cones in $\mathcal{E}$ "get steeper" as p "gets higher".

Figure 3.4

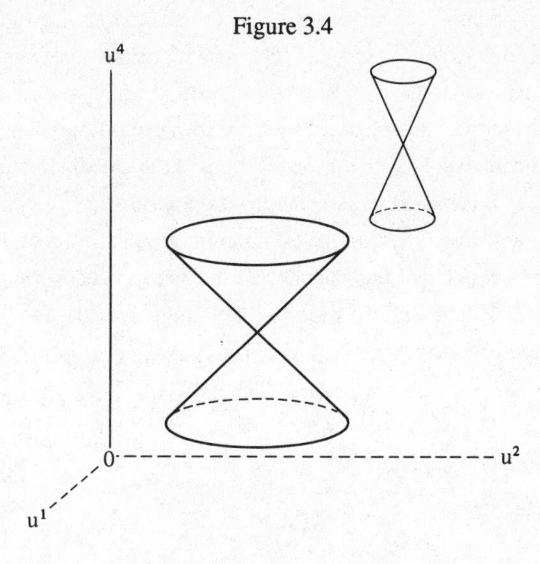

A very useful procedure for producing new spacetimes from old is to observe that any connected open subset M' of a spacetime M is again a spacetime with the obvious Lorentz metric it inherits from M (note that $T_p(M') = T_p(M)$ for each p in M'). Many useful examples can be constructed from known spacetimes by deleting judiciously selected closed subsets. Such an open submanifold of $\mathcal{D}$ was used by Bondi and Gold [BoG] and Hoyle [Ho] as a model of their "steady state" universe and we shall find in section 3.5 that $M = \mathcal{M} - \{(0,0,0,0)\}$ is the simplest example of a spacetime which fails to be "globally hyperbolic".

Different spacetimes are intended to model different physical situations. In classical general relativity the manifold of interest was constructed from specific assumptions about the nature of the gravitational field being modeled by solving Einstein's "field equations" which relate the Lorentz metric to the total energy-momentum tensor (see section 3.6). Minkowski spacetime emerges as the solution appropriate for modeling the event world when all gravitational fields are negligible. The deSitter spacetime can be regarded either as the solution corresponding to a (physically unrealistic) negative energy density or as the "empty universe" solution to a certain modified version of the field equations (with "nonzero cosmological constant"). Einstein-deSitter spacetime is the simplest of all the cosmological models in relativity and represents (to a first approximation) the event world in the presence of a gravitational field due to a uniform ("homogeneous" and "isotropic") "dust" of galaxies in our universe (see Problem 2.F).

The Einstein field equations constitute a very complex system of partial differential equations which can be solved explicitly only by making a great many physically unrealistic symmetry assumptions. As a result any physical implications drawn from such solutions are, at least to this extent, suspect. "Topological" results in general relativity are those which do not depend on the specific form of the field equations, but essentially only on the fact that the event world is modeled by a 4-dimensional Lorentz manifold. Rather than the local coordinate expressions for the metric it is the global structure of the manifold that is of interest. It is indeed remarkable that such results (divorced as they are from the details of the metric structure) can have important physical content, but they do and we shall devote the remainder of this chapter and the next to uncovering some of them. Our investigation will culminate in a proof of the simplest of the famous "Singularity Theorems" of Stephen Hawking.

3.3. Time Orientability, Geodesics and the Chronology Relation

The most serious obstacle to a general, global study of spacetime manifolds is their overwhelming number and diversity. Smooth 4-manifolds are, to say the least, plentiful and "almost" all of them admit Lorentz metrics.* Most of these are, however, physically meaningless so we must begin by trying to narrow the field of view somewhat. We shall impose certain additional restrictions on our Lorentz 4-manifolds which any "reasonable" model of the event world should satisfy.

A smooth curve $\alpha: I \to M$ in a spacetime M is *spacelike, timelike* or *null* if its velocity vector $\alpha'(t)$ is spacelike, timelike or null respectively for each t in I. As in special relativity, a timelike curve is identified with the worldline of a material particle, but in the context of an arbitrary spacetime we face a problem which, in $\mathcal{M}$, was rendered trivial by the existence of a global coordinate system. A material particle's worldline is, after all, more than just a curve. It is a curve traversed in a specific "direction" (from "past" to "future") determined by what has been called the "arrow of time". Now at any fixed point p of M the distinction between "past-directed" and "future-directed" timelike tangent vectors is as easy to make as it is in $\mathcal{M}$. Indeed, if $\{e_i\}_{i=1}^4$ is any orthonormal basis for $T_p(M)$ and $\mathrm{v} = \mathrm{v}^i e_i$ and $w = w^i e_i$ are both timelike, then $\mathrm{v}^4 w^4 > 0$ iff $g(\mathrm{v}, w) < 0$ (Theorem 1.3.1). Consequently, the relation $\mathrm{v} \sim w$ iff $g(\mathrm{v}, w) < 0$ is an equivalence relation on the set of timelike vectors in $T_p(M)$ with precisely two equivalence classes (the two "time cones" at p). One can then designate the elements in one of these two classes as "future-directed" and those in the other "past-directed" (the choice is quite arbitrary, of course, just as the choice of a "standard" basis for $\mathcal{M}$ is arbitrary). To provide an orientation in time along the entire length of a curve, however, this local distinction between past and future at each point is not enough. The choices made must be *consistent*, i.e., they should vary smoothly from point to point along the curve and, should the curve be closed, we must insist that upon traversing it and returning to our point of departure we do not find ourvelves in a different equivalence class. The situation here is not unlike that of the Möbius strip in $\mathbb{R}^3$ which has the property that there is no consistent, smooth choice of a normal vector over the entire surface even though normal vectors are easily selected at each point (see Figure 3.5). One can equally well imagine a spacetime in which any smooth selection of equivalence classes to designate future-directed must "turn upon itself" and be inconsistent.

---

* A Lorentz metric can be defined on *any* noncompact 4-manifold; a compact 4-manifold admits a Lorentz metric if and only if its Euler characteristic is zero (see [O2]).

Figure 3.5

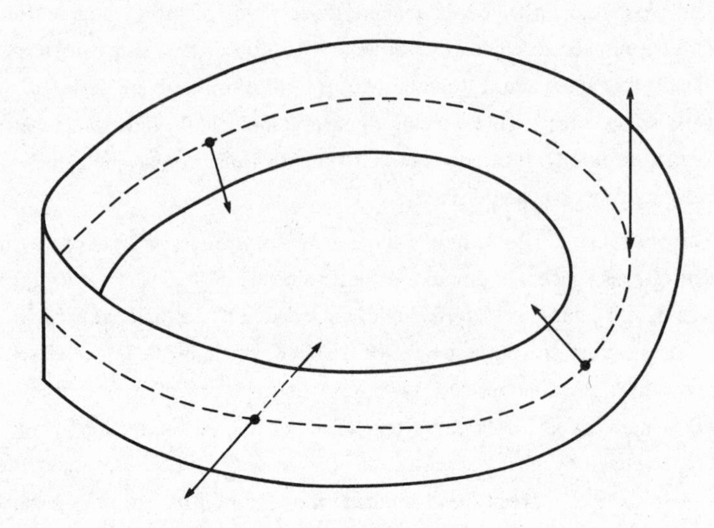

To prohibit such anomolous behavior we shall insist that our spacetimes all satisfy an "orientability" condition which we arrive at as follows: First observe that selecting one of the two equivalence classes of timelike vectors at each p in M to designate future-directed is equivalent to selecting, at each p, just one timelike vector $V(p)$ and stipulating that another timelike v in $T_p(M)$ is future-directed iff $g(V(p), \mathrm{v}) < 0$. Thus we define a *vector field* on an arbitrary manifold M to be a function V which associates with each p in M a unique tangent vector $V(p)$ in $T_p(M)$. V is said to be *smooth* if, for any coordinate patch χ, the *component functions* V^i, $i = 1,...,k$ defined by $V(p) = V^i(p)\,\chi_i(p)$ are C^∞ on $\chi(D)$.

Exercise 3.3.1. Show that, despite appearances, this definition of smoothness does not depend on the choice of χ, i.e., that if $\bar{\chi}: \bar{D} \to M$ is another coordinate patch with $V(p) = \bar{V}^i(p)\,\bar{\chi}_i(p)$ on $\bar{\chi}(\bar{D})$ and if $\chi(D) \cap \bar{\chi}(\bar{D}) \neq \varnothing$, then, on this intersection the V^i are C^∞ if and only if the $\bar{V}_i$ are C^∞. **Hint.** Use Exercise 3.2.16.

Now we shall say that a spacetime M is *time orientable* if one can define on it a smooth vector field V which is everywhere timelike. Henceforth we will assume that our spacetimes are *time oriented*, i.e., time orientable with a specific choice of some smooth timelike vector field V (on $\mathcal{M}$, $\mathcal{D}$ and $\mathcal{E}$ we take $V = \chi_4$, where χ is the standard coordinate patch). One then says that a timelike tangent vector v at some point p of M is *future-*

directed if $g\,(V(p), v) < 0$ and *past-directed* otherwise. A timelike curve $\alpha: I \to M$ is *future-directed* if $g\,(V(\alpha(t)), \alpha'(t)) < 0$ for all t in I and *past-directed* otherwise. Future-directed timelike curves are identified with *worldlines of material particles* in M.

Let $\alpha: I \to M$ be a continuous curve in the manifold M and set $a = \inf I$ and $b = \sup I$ (perhaps $a = -\infty$ and / or $b = \infty$). A point $p \in M$ is called an *endpoint* of α if either $\lim_{t \to a^+} \alpha(t) = p$ or $\lim_{t \to b^-} \alpha(t) = p$. A curve with no endpoints is said to be *endless*. If α is a future-directed timelike curve in a spacetime and $\lim_{t \to a^+} \alpha(t) = p$, then p is a *past endpoint* of α; *future endpoint* is defined similarly. Such an α is *future* (resp., *past*) *endless* if it has no future (resp., past) endpoint. Whether or not a curve has endpoints depends on the manifold it is thought of as living in, e.g., in $\mathcal{M}$ the curve $\alpha(t) = (0,0,0,t)$, $o < t < \infty$ has the origin as its past endpoint, but in $M = \mathcal{M} - \{(0,0,0,0)\}$ the same curve is past endless. As a matter of convenience *we will insist that a smooth timelike curve contain any endpoints it might have* (and so, in particular, must be smooth and timelike at these endpoints). We thereby eliminate from consideration timelike curves which do various physically unreasonable things (e.g., "wiggle" so violently that they fail to have a well-defined tangent at some point, "become null" at an endpoint, etc.)

Let us think for a moment about the sort of curve in a spacetime M which should model the worldline of a material particle which is "free", i.e., influenced only by the gravitational field being modeled by M ("free" = "in free fall"). At each point on such a worldline there exists a local inertial frame as described in Section 3.1 (a nearby freely falling elevator) and relative to such a frame the particle's worldline will appear "straight" since it too is in free fall. The worldline is "locally straight", i.e., it bends only as much as it must to respond to the given gravitational field. In a sense, it bends only as much as it must to "remain in M". Curves on the 2-sphere with this property are the (constant speed) great circles. In an arbitrary Lorentz (or Riemannian) manifold they are called "geodesics". We formulate a definition directly analogous to that for surfaces such as S^2 in $\mathbb{R}^3$.

Exercise 3.3.2. Consider a smooth surface (i.e., 2-manifold) M in $\mathbb{R}^3$ with Riemannian metric g obtained by restricting the $\mathbb{R}^3$-inner product to each tangent plane. Let $\chi: D \to M$ be a coordinate patch for M and $\alpha = \alpha(t)$ a smooth curve in M whose image is contained in $\chi(D)$. Then α's acceleration α'' will, in general, resolve into both tangential and normal components $\alpha'' = \alpha''_{\tan} + \alpha''_{\text{nor}}$, where $\alpha''_{\tan} \in T_{\alpha(t)}(M)$ and $\alpha''_{\tan} \cdot \alpha''_{\text{nor}} = 0$ in $\mathbb{R}^3$.

(a) Write $\alpha(t) = \chi\,(u^1(t), u^2(t))$ and show that

$$\alpha'' = (u^i)''\chi_i + (u^i)'(u^j)'\,\chi_{ij}\,,$$

where $\chi_{ij} = \dfrac{\partial}{\partial u^j}\,\chi_i = \left(\dfrac{\partial^2 x^1}{\partial u^j \partial u^i}, \dfrac{\partial^2 x^2}{\partial u^j \partial u^i}, \dfrac{\partial^2 x^3}{\partial u^j \partial u^i}\right)$.

(b) Resolve χ_{ij} into tangential and normal components to show that

$$\chi_{ij} = \Gamma^r_{ij}\,\chi_r + L_{ij} U\,,$$

where $U = \chi_1 \times \chi_2 / \|\chi_1 \times \chi_2\|$ is the unit normal to M, $L_{ij} = \chi_{ij} \bullet U$ and $\chi_{ij} \bullet \chi_k = \Gamma^r_{ij}\, g_{rk}$.

(c) Define $\Gamma_{r,ij} = g_{rl}\,\Gamma^l_{ij}$ and show that

$$\frac{\partial g_{ij}}{\partial u^k} = \Gamma_{i,jk} + \Gamma_{j,ik}\,. \tag{5}$$

(d) Show that

$$\Gamma^r_{ij} = \frac{1}{2} g^{rk}\left(\frac{\partial g_{ik}}{\partial u^j} + \frac{\partial g_{jk}}{\partial u^i} - \frac{\partial g_{ij}}{\partial u^k}\right). \tag{6}$$

Hint. Permute the indices ijk in (5) to obtain expressions for each of the derivatives in (6) and combine using the symmetries $\Gamma_{i,jk} = \Gamma_{i,kj}$.

(e) Conclude that $\alpha''_{\text{nor}} = L_{ij}(u^i)'(u^j)'U$ and $\alpha''_{\text{tan}} = ((u^r)'' + \Gamma^r_{ij}(u^i)'(u^j)')\,\chi_r$, where the "Christoffel symbols" Γ^r_{ij} are given by (6).

Remarks. If, in Exercise 3.3.2, $t = s$ is the arc length parameter for α, then $\alpha'' = \alpha''(s)$ is the curvature vector of α. One regards α''_{nor} as that part of α's curvature which it must possess simply because it is constrained to lie in M and α''_{tan} the part which α contributes on its own by curving "in M". Curves for which $\alpha''_{\text{tan}} = 0$, i.e., curves $\alpha = \alpha(s) = \chi(u^1(s), u^2(s))$ parametrized by arc length that satisfy

$$(u^r)'' + \Gamma^r_{ij}\,(u^i)'(u^j)' = 0,\ \ r = 1,2,$$

are thought to curve only as much as they must to remain in M. They are the closest thing in M to a straight line and are called the "geodesics" of the surface, e.g., if M is a sphere, they are the constant speed parametrizations of the great circles (see Problem 3.A).

Now we let M be an arbitrary manifold with metric and define, for each coordinate patch χ on M, the *Christoffel symbols* (of the second kind) Γ^r_{ij} by

$$\Gamma^r_{ij} = \frac{1}{2} g^{rl} \left(\frac{\partial g_{il}}{\partial u^j} + \frac{\partial g_{jl}}{\partial u^i} - \frac{\partial g_{ij}}{\partial u^l} \right), \quad r,i,j = 1,\ldots,k \ ,$$

where g_{ab} are the metric components relative to χ, (g^{ab}) is the matrix inverse of (g_{ab}), $u^1,\ldots,u^k$ are the Cartesian coordinates in the domain of χ and the summation convention applies over $l = 1,\ldots,k$. A smooth curve $\mu = \mu(t)$ is said to be a *geodesic* of M if, for each χ,

$$(u^r)'' + \Gamma^r_{ij}(u^i)'(u^j)' = 0, \ r = 1,\ldots,k \tag{7}$$

where $\mu(t) = \chi(u^1(t),\ldots,u^k(t))$.

Exercise 3.3.3. Let $\chi\colon D \to M$ and $\bar{\chi}\colon\bar{D} \to M$ be two coordinate patches for M with $\bar{\chi}(\bar{D}) \cap \chi(D) \neq \varnothing$. Prove that, where the two coordinate systems overlap,

$$\bar{\Gamma}^r_{ij} = \frac{\partial \bar{u}^r}{\partial u^s} \frac{\partial u^m}{\partial \bar{u}^i} \frac{\partial u^l}{\partial \bar{u}^j} \Gamma^s_{ml} + \frac{\partial \bar{u}^r}{\partial u^s} \frac{\partial^2 u^s}{\partial \bar{u}^i \partial \bar{u}^j}$$

and use this to show that the defining equations (7) for a geodesic are satisfied in χ iff they are satisfied in $\bar{\chi}$.

Remark. Henceforth we shall leave it as a *STANDING EXERCISE* to verify that definitions which appear to depend on the choice of coordinate system, in fact, do not.

Exercise 3.3.4. Show that the only nonzero Christoffel symbols for the standard coordinate patch on $\mathcal{E}$ are

$$\Gamma^1_{14} = \Gamma^1_{41} = \Gamma^2_{24} = \Gamma^2_{42} = \Gamma^3_{34} = \Gamma^3_{43} = \frac{2}{3}(u^4)^{-1} \ .$$

Remark. Observe that (7) is trivially satisfied by any "constant curve" ($\mu(t) = p$ for each t). While such a curve has zero velocity vector and therefore technically does not qualify as a smooth curve according to our definition, it will be convenient to bend our chosen terminology somewhat and refer to such constant curves as *degenerate geodesics.*

If one is confronted with a manifold and assigned the task of finding its geodesics, the usual procedure is to fix a coordinate patch, calculate the metric components g_{ij} and the Christoffel symbols Γ^r_{ij}, substitute into (7) and solve the resulting system of (second order, nonlinear) ordinary differential equations for the component functions

$u^r(t)$, $r = 1,...,k$. With few exceptions this is an arduous task. One of the exceptions, of course, is Minkowski spacetime where, relative to the usual coordinate patch, the metric components are constant so that the Christoffel symbols are all zero. The geodesic equations (7) therefore become $(u^r)'' = 0$, $r = 1,2,3,4$ so that $u^r(t) = A_r t + B_r$, where A_r and B_r are constants. The geodesics of $\mathcal{M}$ are, as expected, linearly parametrized straight lines.

We shall exhibit some geodesics of $\mathcal{D}$ and $\mathcal{E}$ shortly, but first we must develop some tools. We focus our attention on a fixed point p in some manifold M with metric and a fixed tangent vector v in $T_p(M)$. We shall say that a smooth curve α in M defined on some interval about zero *fits* v *at* p if it goes through p with velocity v at $t = 0$, i.e., if

$$\alpha(0) = p \quad \text{and} \quad \alpha'(0) = \text{v} \ . \tag{8}$$

Appealing to the standard Existence and Uniqueness Theorem for solutions to initial value problems for systems such as (7) (see [Har]) we find that for every p in M and every v in $T_p(M)$ there exists a geodesic of M that fits v at p and, assuming that its domain has been extended to the largest possible interval about zero, that there will be only one such (all of this is, of course, obvious in $\mathcal{M}$). More precisely we have

Theorem 3.3.1. Let p be a point in the (Riemannian or Lorentzian) manifold M and v a tangent vector to M to p. Then there exists a unique maximal geodesic μ defined on some interval I about zero which fits v at p ("maximal" means that if $\lambda: J \to M$ is another geodesic that fits v at p, then $J \subseteq I$ and $\lambda = \mu \mid J$).

Remark. If $\text{v} = 0$ we take μ to be the degenerate geodesic $\mu(t) = p$ for $-\infty < t < \infty$.

The uniqueness assertion in Theorem 3.3.1 is often useful for finding the geodesics in a given manifold since it assures us that if we have somehow managed to conjure up geodesics in every "direction" v at p, then we will, in fact, have all of the geodesics through p. We apply this procedure to the deSitter spacetime $\mathcal{D}$. We will show that, because $\mathcal{D}$ is a level hypersurface in $\mathbb{R}^5_1$ and its metric is the restriction of the "Minkowski inner product" on $\mathbb{R}^5_1$, Exercise 3.3.2 for surfaces in $\mathbb{R}^3$ goes through essentially verbatim for $\mathcal{D}$. First, however, we must define a "normal vector" to $\mathcal{D}$ in $\mathbb{R}^5_1$ and for this we introduce a few new ideas that will be required in other contexts as well.

We consider an arbitrary smooth real-valued function f on some manifold M. For each p in M and every v in $T_p(M)$ we let v$[f]$ denote the derivative of f in the direction v at p, i.e.,

$$\mathrm{v}[f] = \frac{d}{dt} f(\alpha(t))\Big|_{t=0} \qquad (9)$$

where $\alpha = \alpha(t)$ is any smooth curve in M which fits v at p. Observe that if χ is any coordinate patch at p in M with $\alpha(t) = \chi(u^1(t),...,u^k(t))$ and $\mathrm{v} = \mathrm{v}^i \chi_i$, then

$$\mathrm{v}[f] = \frac{\partial f}{\partial u^i}(p) \frac{du^i}{dt}(0) = \frac{\partial f}{\partial u^i} v^i \qquad (10)$$

so that the definition clearly does not depend on the choice of α (we shall permit ourselves such minor and obvious abuses of notation as writing $\frac{\partial f}{\partial u^i}(p)$ for $\frac{\partial (f \circ \chi)}{\partial u^i}(\chi^{-1}(p)))$. Next we define a smooth vector field on M called the *gradient of f* and denoted ∇f. Near each point p of M select a coordinate patch χ and set

$$\nabla f = g^{ij} \frac{\partial f}{\partial u^i} \chi_j$$

(Don't forget your "Standing Exercise" from page 95). For example, if $Q : \mathcal{M} \to \mathbb{R}$ is defined as usual by $Q(u^1,u^2,u^3,u^4) = (u^1)^2 + (u^2)^2 + (u^3)^2 - (u^4)^2$, then $\nabla Q = g^{ij} \frac{\partial Q}{\partial u^i} \chi_j = \eta^{ij} \frac{\partial Q}{\partial u^i} e_i = 2u^1 e_1 + 2u^2 e_2 + 2u^3 e_3 + 2u^4 e_4 = 2u^i e_i$. Thus, the gradient of the "squared length" function in $\mathcal{M}$ is *radial.* i.e., a multiple of the "position vector" $u^i e_i$.

Exercise 3.3.5. Let $T : \mathcal{E} \to \mathbb{R}$ be the fourth coordinate function of the usual coordinate patch on Einstein-deSitter spacetime, i.e., $T(u^1,u^2,u^3,u^4) = u^4$. Show that (g^{ij}) is a diagonal matrix with $g^{ii} = 1/g_{ii}$ for $i = 1,2,3,4$, and then that $\nabla T = -e_4$.

The gradient has all of the properties familiar from vector calculus, e.g., the derivative of f in the direction v at p can be computed by "dotting" v and $\nabla f(p)$:

Lemma 3.3.2. Let M be a smooth manifold with metric, $p \in M$, $\mathrm{v} \in T_p(M)$ and $f : M \to \mathbb{R}$ a smooth function. Then

$$\mathrm{v}[f] = \langle \nabla f(p), \mathrm{v} \rangle \; . \qquad (11)$$

Proof: Choose a smooth curve α which fits v at p. Then, for any coordinate patch χ, $\langle \nabla f(p), \mathrm{v} \rangle = \langle \nabla f(p), \alpha'(0) \rangle = g_{ab} \left(g^{ia} \frac{\partial f}{\partial u^i}(p) \frac{du^b}{dt}(0) \right) = \delta^i_b \frac{\partial f}{\partial u^i}(p) \frac{du^b}{dt}(0) =$

v[f] by (10). Q.E.D.

Exercise 3.3.6. Show that $\nabla f(p) = 0$ if and only if $\dfrac{\partial f}{\partial u^i}(p) = 0$, $i = 1,\ldots,k$, for every coordinate patch χ at p.

In section 5 of this chapter we will generalize Exercise 3.2.8 and show that if M is a smooth k-manifold, $f: M \to \mathbb{R}$ is a smooth map and $r \in \mathbb{R}$ is one its values, then $S = \{p \in f^{-1}(r) \colon \nabla f(p) \neq 0\}$ is a smooth $(k-1)$-dimensional submanifold of M. Since f is constant along any smooth curve in S it will follow from Lemma 3.3.2 that ∇f is orthogonal in M to every $T_p(S)$. Here we will require only the following special case: Let $Q_1 \colon \mathbb{R}^5 \to \mathbb{R}$ be defined by $Q_1(x) = g_1(x,x) = (x^1)^2 + (x^2)^2 + (x^3)^2 + (x^4)^2 - (x^5)^2$. Then $\nabla Q_1(x) = 2x$. Now, we know that $\mathcal{D} = Q_1^{-1}(1)$ is a smooth 4-dimensional submanifold of $\mathbb{R}^5$. Since every tangent vector to $\mathcal{D}$ is the velocity vector of some smooth curve in $\mathcal{D}$ and since Q_1 is constant along any such curve, Lemma 3.3.2 implies that the $\mathbb{R}^5_1$-inner product of ∇Q_1 with any tangent vector to $\mathcal{D}$ is zero, i.e., ∇Q_1 is "normal" to $\mathcal{D}$ in $\mathbb{R}^5_1$. Moreover, for each x in $\mathcal{D}$, $<\nabla Q_1(x), \nabla Q_1(x)> = <2x, 2x> = 4<x,x> = 4$ so $U(x) = \frac{1}{4}\nabla Q_1(x)$ defines a *unit normal* vector field to $\mathcal{D}$ in $\mathbb{R}^5_1$.

Exercise 3.3.7. Repeat the arguments in Exercise 3.2.8 (using $U = \frac{1}{4}\nabla Q_1$ as the unit normal to $\mathcal{D}$ in $\mathbb{R}^5_1$) to show that the acceleration α'' of any smooth curve in $\mathcal{D}$ is given by

$$\alpha'' = ((u^r)'' + \Gamma^r_{ij}(u^i)'(u^j)')\,\chi_r + (L_{ij}(u^i)'(u^j)')\,U \ . \tag{12}$$

We conclude then that *a smooth curve in $\mathcal{D}$ is a geodesic of $\mathcal{D}$ iff its acceleration is normal to $\mathcal{D}$ in $\mathbb{R}^5_1$.* Since the normal direction to $\mathcal{D}$ in $\mathbb{R}^5_1$ is "radial" ($\nabla Q_1(x) = 2x$) it is rather easy to geometrically "guess" the images of these geodesics. We also "guess" the appropriate parametrizations by anticipating a result proved later (Corollary 3.3.5) according to which any geodesic is a constant speed curve, i.e., has $g(\mu'(t), \mu'(t))$ constant.

Figure 3.6

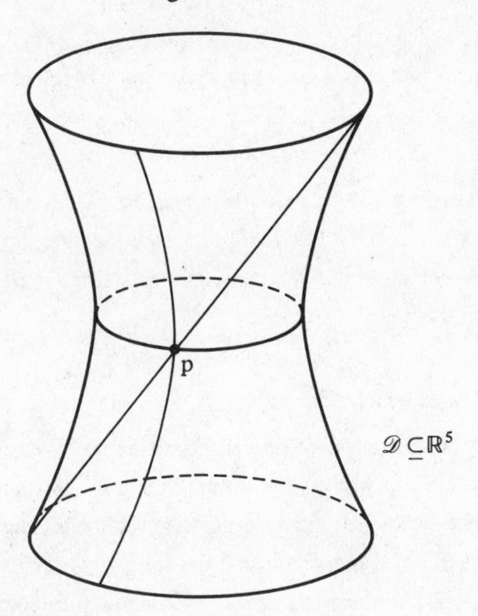

Now fix a p in $\mathcal{D}$ and a nonzero v in $T_p(\mathcal{D})$. We determine the unique geodesic of $\mathcal{D}$ that fits v at p. Since the normal to $\mathcal{D}$ is radial the vectors p and v in $\mathbb{R}^5_1$ are orthogonal. These vectors determine a unique 2-dimensional plane $\mathcal{P}$ in $\mathbb{R}^5$. Since p is "spacelike" in $\mathbb{R}^5_1$ there are three possibilities:

1. v is spacelike so that $e_1 = v/<v,v>^{1/2}$ is a unit spacelike vector in $\mathbb{R}^5_1$ and $\{e_1, p\}$ is an orthonormal basis for $\mathcal{P}$. In this case a point $ae_1 + bp$ of $\mathcal{P}$ is in $\mathcal{D}$ iff $a^2 + b^2 = 1$. Thus, $\mathcal{P} \cap \mathcal{D}$ is the unit circle in $\mathcal{P}$ which can be parametrized as

$$\mu(t) = (\sin kt)e_1 + (\cos kt)p, \ -\infty < t < \infty \ ,$$

where $k = <\text{v},\text{v}>^{1/2}$. Then $\mu''(t) = -k^2\mu(t)$ so μ'' is everywhere normal to $\mathcal{D}$ in $\mathbb{R}^5_1$, i.e., μ is a geodesic of $\mathcal{D}$. Moreover, $\mu(0) = p$ and $\mu'(0) = ke_1 = v$. Observe that $g(\mu'(t), \mu'(t)) = k^2 = <v,v>$ so μ is everywhere spacelike.

2. v is timelike. Let $e_2 = \text{v}/(-<\text{v},\text{v}>)^{1/2}$. Then $\{e_2, p\}$ is an orthonormal basis for $\mathcal{P}$ with $g(e_2, e_2) = -1$. Now a point $ae_2 + bp$ of $\mathcal{P}$ is in $\mathcal{D}$ iff $-a^2 + b^2 = 1$ so $\mathcal{P} \cap \mathcal{D}$ consists of two branches of a hyperbola in $\mathcal{P}$. The branch through p can be parametrized as

$$\mu(t) = (\sinh kt)e_2 + (\cosh kt)p, \ -\infty < t < \infty \ ,$$

where $k = (-\langle v,v \rangle)^{1/2}$.

Exercise 3.3.8. Show that μ is a geodesic which fits v at p and is, moreover, everywhere timelike.

3. v is null (and nonzero). Thus, $\{v,p\}$ is a basis for $\mathcal{P}$ and a point $a\mathrm{v} + bp$ of $\mathcal{P}$ is in $\mathcal{D}$ iff $a^2 \cdot 0 + b^2 \cdot 1 = 1$, i.e., $b^2 = 1$ so $b = \pm 1$. Consequently, $\mathcal{P} \cap \mathcal{D}$ consists of two parallel straight lines, the one through p having parametrization

$$\mu(t) = p + t\mathrm{v}, \quad -\infty < t < \infty .$$

μ is then a null geodesic which fits v at p.

The geodesics of Einstein-deSitter spacetime are not so easily identified, but here are some examples (everything we say, but do not verify for $\mathcal{E}$ is proved in detail in [SW2]; nevertheless, it would be an instructive exercise to write out the geodesic equations for the usual coordinate patch on $\mathcal{E}$ and verify that the following curves are solutions). For convenience we shall omit reference to $\chi = id_{R^4}$ and denote the coordinates in $\mathcal{E}$ by u^1, u^2, u^3 and u^4. The vertical straight lines $\mu(t) = (u^1(t), u^2(t), u^3(t), u^4(t)) = (u_0^1, u_0^2, u_0^3, t)$, $-\infty < t < \infty$, are timelike geodesics in $\mathcal{E}$.

Remark. These vertical geodesics are identified with the worldlines of the galaxies (or galaxy clusters) of our universe. The "displacement vector" between two events with the same u^4 on two different vertical geodesics is spacelike and satisfies $\| V \| = (g(V,V))^{1/2} = K(u^4)^{2/3}$, where K is a positive constant (here we have in mind two "nearby" galaxies so that the vector V, which actually lies in a tangent space to $\mathcal{E}$, can be thought of as joining the galaxies). Observe that $\| V \|$, which is regarded as the distance between the two galaxies at the "instant" u^4 in the given global coordinate system, satisfies $\frac{d}{du^4} \| V \| = \frac{2}{3} K (u^4)^{-1/3} > 0$ and $\frac{d^2}{d(u^4)^2} \| V \| = -\frac{2}{9} K (u^4)^{-4/3} < 0$ so that, in this particular cosmological model, the galaxies are receding from each other (i.e., the universe is expanding), but at a decreasing rate.

The cubic curve $\lambda(t) = (0, 3t^{1/5}, 0, t^{3/5})$, $-\infty < t < \infty$, is a null geodesic in $\mathcal{E}$. Indeed, one can show (see pages 133-135 and 161-162 of [SW2]) that *any* null geodesic in $\mathcal{E}$ can be obtained from this "standard photon" by reparametrizing affinely (i.e., introducing a new parameter s defined by $t = ms + b$, where m and b are constants) and/or composing with some map $H: \mathcal{E} \to \mathcal{E}$ of the form $H(\vec{w}, t) = (h(\vec{w}), t)$, where h is an isometry of $\mathbb{R}^3$

onto itself (i.e., a rotation, translation or reflection).

Figure 3.7

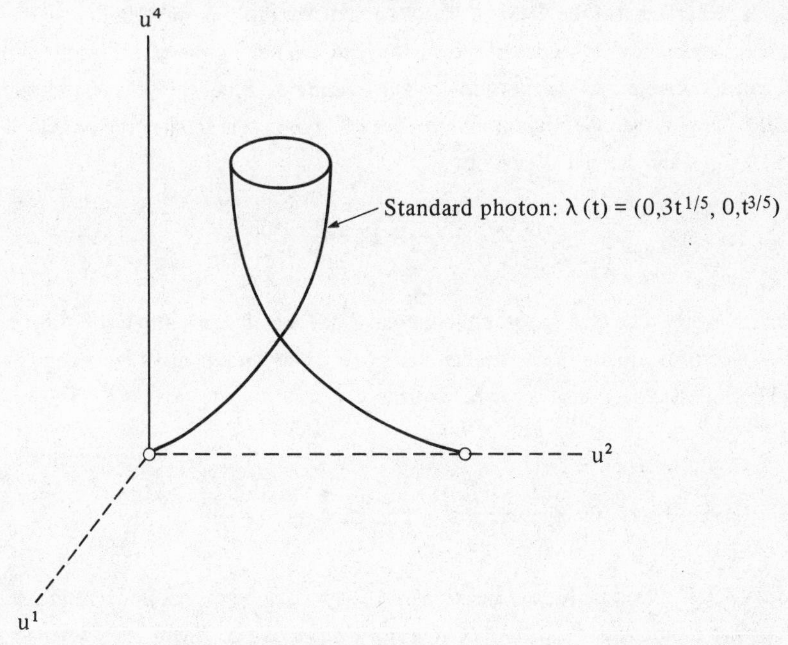

For our study of chronology in spacetimes we shall require a number of basic properties of geodesics, but before deriving them it will be convenient to rephrase the definition along more contemporary lines and introduce some standard terminology and notation. Let $\alpha: I \to M$ be an arbitrary smooth curve in the manifold M. A *vector field along* α is a function V which assigns to each t in I a tangent vector $V(t)$ in $T_{\alpha(t)}(M)$. Observe that for each fixed t_0 in I it is possible, since α is continuous, to select a subinterval J of I with t_0 in its interior such that $\alpha(J)$ is contained in $\chi(D)$ for some coordinate patch χ. We will say that V is *smooth at* t_0 if, for each such choice of J and χ, the component functions $V^i(t)$ defined by $V(t) = V^i(t)\,\chi_i(\alpha(t))$, $t \in J$, are C^∞. V is *smooth* if it is smooth at each t_0 in I. The most obvious example, of course, is the velocity vector field $t \to \alpha'(t)$. A smooth vector field V along α is said to be *parallel along* α if it does not change "tangentially" along α, if for each χ,

$$\frac{dV^r}{dt} + \Gamma^r_{ij}\frac{du^j}{dt}V^i = 0, \quad r = 1,\ldots,k \;, \tag{13}$$

where $\alpha(t) = \chi(u^1(t),\ldots,u^k(t))$. Thus, a geodesic is a curve μ whose velocity vector field μ' is parallel along μ (there is an interesting discussion of the physical and geometrical motivation for this definition in [ABS]). Observe that if α is given, then (13) can be regarded as a linear system of first order ordinary differential equations for the $V^r(t)$. It therefore follows from the basic Existence and Uniqueness Theorem for such systems (see, e.g., [Har]) that if the value of V is specified arbitrarily at some initial t_0, then (13) uniquely determines the values of the vector field $V(t)$ along the entire length of α. We shall call $V(t)$ the *parallel translation of* $V(t_0)$ *along* α. Finally, if V is any smooth vector field along α, then the *covariant derivative of* V *along* α is another vector field along α denoted $D_{\alpha'}V$ and defined at each t by

$$D_{\alpha'}V = \left[\frac{dV^r}{dt} + \Gamma^r_{ij}\frac{du^j}{dt}V^i\right]\chi_r \,, \tag{14}$$

where t is again restricted to some subinterval J of I which maps into a coordinate patch. Thus, a vector field along α is parallel along α if its covariant derivative along α is zero and the geodesic equations (7) can be written

$$D_{\mu'}\,\mu' = 0 \,. \tag{15}$$

When $V = \alpha'$, $D_{\alpha'}\,\alpha'$ is a measure of α's "non-geodesy".

Lemma 3.3.3. (Product Rule): Let $\alpha: I \to M$, $\alpha = \alpha(t)$, be a smooth curve in the manifold M with metric and V and W smooth vector fields along α. Then

$$\frac{d}{dt}(V \bullet W) = V \bullet D_{\alpha'}W + W \bullet D_{\alpha'}V \,. \tag{16}$$

Proof: It will clearly suffice to focus our attention on t's in a subinterval J of I which α maps into $\chi(D)$ for some coordinate patch χ. Writing $\alpha(t) = \chi(u^1(t),...,u^k(t))$, $V = V^r(t)\chi_r(\alpha(t))$ and $W = W^r(t)\chi_r(\alpha(t))$ for t in J we have

$$V \bullet D_{\alpha'}W + W \bullet D_{\alpha'}V = g_{rl}\,V^r\left(\frac{dW^l}{dt} + \Gamma^l_{ij}\frac{du^j}{dt}W^i\right) + g_{rl}W^r\left(\frac{dV^l}{dt} + \Gamma^l_{ij}\frac{du^j}{dt}V^i\right)$$

$$= g_{rl}\left(V^r\frac{dW^l}{dt} + W^r\frac{dV^l}{dt}\right) + g_{rl}\Gamma^l_{ij}\frac{du^j}{dt}\left(V^rW^i + W^rV^i\right)$$

We simplify the second term as follows:

$$g_{rl}\Gamma^l_{ij} = \frac{1}{2}g_{rl}g^{lm}\left(\frac{\partial g_{im}}{\partial u^j} + \frac{\partial g_{jm}}{\partial u^i} - \frac{\partial g_{ij}}{\partial u^m}\right)$$

$$= \frac{1}{2}\delta^m_r\left(\frac{\partial g_{im}}{\partial u^j} + \frac{\partial g_{jm}}{\partial u^i} - \frac{\partial g_{ij}}{\partial u^m}\right)$$

$$= \frac{1}{2}\left(\frac{\partial g_{ir}}{\partial u^j} + \frac{\partial g_{jr}}{\partial u^i} - \frac{\partial g_{ij}}{\partial u^r}\right).$$

Consequently,

$$g_{rl}\Gamma^l_{ij}\frac{du^j}{dt}(V^r W^i + W^r V^i) = \frac{1}{2}\left(\frac{\partial g_{ir}}{\partial u^j} + \frac{\partial g_{jr}}{\partial u^i} - \frac{\partial g_{ij}}{\partial u^r}\right)\frac{du^j}{dt}(V^r W^i + W^r V^i)$$

$$= \frac{1}{2}\frac{\partial g_{ir}}{\partial u^j}\frac{du^j}{dt}(V^r W^i + W^r V^i) +$$

$$+ \frac{1}{2}\left(\frac{\partial g_{jr}}{\partial u^i} - \frac{\partial g_{ij}}{\partial u^r}\right)\frac{du^j}{dt}(V^r W^i + W^r V^i)$$

$$= \frac{\partial g_{ir}}{\partial u^j}\frac{du^j}{dt}V^i W^r$$ since the first term is symmetric and the second

is skew symmetric in i and r

$$= \frac{dg_{ir}}{dt}V^i W^r = \frac{dg_{rl}}{dt}V^r W^l .$$

Thus we find that

$$V \bullet D_{\alpha'}W + W \bullet D_{\alpha'}V = g_{rl}\left(V^r\frac{dW^l}{dt} + W^r\frac{dV^l}{dt}\right) + \frac{dg_{rl}}{dt}V^r W^l$$

$$= g_{rl}V^r\frac{dW^l}{dt} + g_{rl}W^l\frac{dV^r}{dt} + \frac{dg_{rl}}{dt}V^r W^l$$

$$= \frac{d}{dt}(g_{rl}V^r W^l) = \frac{d}{dt}(V \bullet W)$$

Q.E.D.

Theorem 3.3.4. Parallel translation along a smooth curve $\alpha: I \to M$ preserves dot products, i.e. if V and W are parallel along α, then $g(V(t), W(t))$ is constant.

Proof: $\frac{d}{dt}g(V(t), W(t)) = \frac{d}{dt}(V \cdot W) = V \cdot D_{\alpha'}W + W \cdot D_{\alpha'}V = V \cdot 0 + W \cdot 0 = 0$.
Q.E.D.

From this it follows at once that if a geodesic is spacelike, timelike or null at one point it must have that same character at every point. Indeed, we have

Corollary 3.3.5. If $\mu : I \to M$ is a geodesic of M, then $g(\mu'(t), \mu'(t))$ is constant.

Whether or not a curve is a geodesic depends as much on its parametrization as on its image. If $\alpha : I \to M$ is an arbitrary smooth curve and $h : J \to I$, $t = h(s)$, is smooth function on the interval J with $h'(s) > 0$ on J, then $\beta = \alpha o h : J \to M$ is a smooth curve called a *reparametrization* of α.

Theorem 3.3.6. Let $\mu : I \to M$ be a nondegenerate geodesic. A reparametrization $\mu o h : J \to M$ of μ is a geodesic iff h has the form $h(s) = ms + b$ for some constants m and b.

Proof: By assumption, $\frac{d^2u^r}{dt^2} + \Gamma^r_{ij}\frac{du^i}{dt}\frac{du^j}{dt} = 0$, $r = 1, \ldots, k$. From the chain rule, $\frac{du^r}{ds} = h'(s)\frac{du^r}{dt}$ and $\frac{d^2u^r}{ds^2} = (h'(s))^2\frac{d^2u^r}{dt^2} + (h''(s))\frac{du^r}{dt}$. Thus,

$$\frac{d^2u^r}{ds^2} + \Gamma^r_{ij}\frac{du^i}{ds}\frac{du^j}{ds} = (h'(s))^2\left[\frac{d^2u^r}{dt^2} + \Gamma^r_{ij}\frac{du^i}{dt}\frac{du^j}{dt}\right] + (h''(s))\frac{du^r}{dt}$$

$$= (h''(s))\frac{du^r}{dt} .$$

Since μ is nondegenerate μ' is nonzero so some $\frac{du^r}{dt}$ must be nonzero at each t. Thus, the reparametrized curve can satisfy the geodesic equations iff $h''(s) = 0$, i.e., $h(s) = ms + b$.
Q.E.D.

If M is a spacetime and $\alpha : I \to M$ is an arbitrary smooth timelike curve ($\alpha = \alpha(t)$) and if one fixes some t_0 in I, then a *proper time parameter* τ along α is defined in the usual way by

$$\tau = \tau(t) = \int_{t_0}^{t} |g(\alpha'(t), \alpha'(t))|^{1/2}\, dt \quad . \tag{17}$$

Since $\frac{d\tau}{dt} = |g(\alpha',\alpha')|^{1/2}$ and $g(\alpha',\alpha')$ is never zero, α can be reparametrized in terms of τ. We prefer to use the same name for this proper time parametrization of α and simply write $\alpha = \alpha(\tau)$. Observe that if μ is a timelike geodesic, then Corollary 3.3.5 implies that the proper time parameter τ along μ satisfies $\tau = K(t - t_0)$ for some positive constant K and so, by Theorem 3.3.6, $\mu = \mu(\tau)$ is also a geodesic.

Remark. We again point out that, just as for surfaces in $\mathbb{R}^3$ (see the Remark following Exercise 3.2.15), one often sees (17) abbreviated in "differential form"

$$d\tau^2 = -g_{ij}du^i du^j \ .$$

Before putting all of this information to use in our study of time orientability and chronology we shall introduce one last technical device by which we can make precise our contention that every event can be viewed from a nearby "local inertial frame". Let M be an arbitrary manifold with metric. Fix a $p \in M$ and let D_p be the set of all vectors v in $T_p(M)$ for which the maximal geodesic μ that fits v at p is defined at least on $[0,1]$. The *exponential map* at p is then the function

$$\exp_p : D_p \to M \ .$$

defined by $\exp_p(\text{v}) = \mu(1)$ for all $\text{v} \in D_p$. Thus, $\exp_p(\text{v})$ is the point in M which is parameter distance 1 along the unique geodesic which fits v at p. Observe that $\exp_p(0) = p$ (degenerate geodesic) and also that D_p is *star-shaped* i.e., $t\text{v} \in D_p$ whenever $\text{v} \in D_p$ and $0 \le t \le 1$.

Exercise 3.3.9. Let $p \in M$ and $\text{v} \in D_p$. Show that for every t in the domain of μ (in particular, for $0 \le t \le 1$), $\exp_p(t\text{v})$ is defined and

$$\exp_p(t\text{v}) = \mu(t) \ . \tag{18}$$

One therefore pictures $\exp_p$ as taking the straight line $t\text{v}$, $t \in$ domain (μ), along v in $T_p(M)$ and smoothing it out along the image of μ in M.

Exercise 3.3.10. Describe D_p, $\exp_p$ and $\exp_p(T_p(\mathcal{M}))$ for each p in $\mathcal{M}$.

$\exp_p$ is really quite a nice mapping. That it is C^∞ on some open neighborhood of 0 in $T_p(M)$ follows from the basic theory of ordinary differential equations (C^∞ dependence of solutions to (7) on initial conditions; see [Har]). By calculating its derivative (Jacobian) and appealing to the Inverse Function Theorem one finds that, in fact, it is a diffeomorphism on some neighborhood of 0, i.e., $\exp_p$ is a coordinate patch at p in M. Cutting down its domain a bit more, this coordinate patch acquires the very useful property of "geodesic convexity":

Theorem 3.3.7. For each p in the manifold M with metric there exists an open, star-shaped neighborhood N of 0 in $T_p(M)$ and an open neighborhood N_p of p in M such that

(a) $\exp_p$ is a diffeomorphism of N onto N_p (such an N_p is called a *normal neighborhood* of p),

(b) N_p is *geodesically convex,* i.e., it is a normal neighborhood of each of its points (for each $q \in N_p$ there exists an open, star-shaped neighborhood of 0 in $T_q(M)$ which $\exp_q$ maps diffeomorphically onto N_p).

The proof (especially of (b), which is due to J.H.C. Whitehead) is rather involved and we shall simply direct the readers attention to the treatments in [Hi], section 9.3 and 9.4 and [O2], pages 70-72 and 129-130. Observe that it follows from Theorem 3.3.7 that for any $p \in M$ the geodesically convex normal neighborhoods of p form a *local base* at p, i.e., any open set V containing p also contains such a neighborhood of p. To see this observe that we may clearly assume V is connected and therefore a spacetime in its own right. Moreover, $T_q(V) = T_q(M)$ for any $q \in V$ and geodesic segments of V are also geodesic segments of M (Why?). Thus, applying Theorem 3.3.7 to V yields a geodesically convex normal neighborhood of p *in* M which is contained in V.

Exercise 3.3.11. Show that any two points in a geodesically convex normal neighborhood N_p in M can be connected by one and only one geodesic segment contained in N_p.

Remarks. Two given points in a manifold M need not be connectible by a geodesic segment, e.g., $(0,0,0,-1)$ and $(0,0,0,1)$ in $\mathcal{M} - \{(0,0,0,0)\}$ and, even if they are, that geodesic segment need not be unique, e.g., any two points on the sphere S^2 can be joined by two arcs of a great circle (find a similar example in deSitter spacetime $\mathcal{D}$). Even two points inside a given normal neighborhood N_p might be connectible by a number of geodesic segments, but only one of them can remain entirely in N_p (Example?).

Now fix a basis $\{e_1, \ldots, e_k\}$ for $T_p(M)$. Each point v in $T_p(M)$ thereby acquires coordinates: $\mathrm{v} = \mathrm{v}^i e_i$. As does any coordinate patch, $\exp_p$ transfers this coordinate system to its image N_p, i.e., for each v in $T_p(M)$, $\exp_p(\mathrm{v})$ is assigned coordinates $(\mathrm{v}^1, \ldots, \mathrm{v}^k)$. Any such coordinate system on M is called a *normal coordinate system* at p. If M is a spacetime and $\{e_1, e_2, e_3, e_4\}$ is an orthonormal basis for $T_p(M)$ with e_4 future-directed timelike, then the corresponding coordinate system at p is called a *Minkowski normal coordinate system* at p and it is these we have in mind when we refer to a "local inertial frame" at p.

At this point we have (finally) assembled enough machinery to return to our study of chronology. Our first objective is to generalize the chronology relation << on $\mathcal{M}$ to an arbitrary spacetime M. Guided by Theorem 1.5.6 one might say that p in M chronologically precedes q in M if there is a smooth, future-directed timelike curve in M from p to q. We formulate our definition in equivalent, but somewhat more convenient terms. A *trip* in M is a continuous curve $\gamma: I \rightarrow M$ which is piecewise a future-directed timelike geodesic, i.e., for which there is a partition of I into subintervals such that the restriction of γ to each subinterval is a future-directed timelike geodesic. As with smooth timelike curves we will require that a trip contain any endpoints it might have. Trips in $\mathcal{M}$ are, of course, just future-directed timelike polygons. A trip with past endpoint p and future endpoint q is a *trip from p to q* and if such a trip exists we shall say that *p chronologically precedes q* and write $p \ll q$. Geodesics are very special, but in some sense, trips are quite general.

Theorem 3.3.8. Let M be a spacetime with p and q in M. Then $p \ll q$ if and only if there exists a smooth, future-directed timelike curve in M from p to q.

If there is a smooth, future-directed timelike curve in M from p to q, then its image is compact and can therefore be covered by a finite "chain" of geodesically convex normal neighborhoods. By convexity one can "hook up" points on the curve in consecutive normal neighborhoods with timelike geodesics and thereby build a trip from p to q. Conversely, if there is a trip from p to q the "joints" can be smoothed to give a timelike curve from p to q by lifting each pair of joined segments to $T_r(M)$ via $\exp_r^{-1}$ (r is the joint), observing that there is enough room between the lifted timelike straight lines and the null cone in $T_r(M)$ to round them off and remain timelike and then pushing back into M via $\exp_r$. The details are available in Proposition 2.23 of [Pen].

Trips are somewhat easier to deal with than smooth timelike curves. For example, the following basic property of the relation $\ll$ is an immediate consequence of our definition.

Lemma 3.3.9. The relation $\ll$ on M is transitive, i.e., if $p \ll q$ and $q \ll r$, then $p \ll r$.

For any p in M we define the *chronological future of p*, denoted $I^+(p)$, by $I^+(p) = \{q \in M : p \ll q\}$. The *chronological past of p* is $I^-(p) = \{q \in M : q \ll p\}$. For any $S \subseteq M$ we let $I^{\pm}(S) = \bigcup_{p \in S} I^{\pm}(p)$. In $\mathcal{M}$, of course, $I^+(p)$ (resp., $I^-(p)$) is just the interior of the upper (resp., lower) null cone at p. Similar statements are true in $\mathcal{D}$ and $\mathcal{E}$, but there is no analogous statement for an arbitrary spacetime. The most that can be said in general is that *locally* (inside a geodesically convex normal neighborhood) a timelike curve cannot escape the null cone (see Lemma 4.5.3). Indeed, chronological past and future sets can be quite pathological and have rather unexpected properties. It can even happen that $p \in I^+(p)$ for some $p \in M$, i.e., that M contains *closed* timelike curves (indeed, if one constructs a spacetime M from the portion of $\mathcal{M}$ between $x^4 = -1$ and $x^4 = 1$ by "identifying" $(x^1,x^2,x^3,-1)$ and $(x^1,x^2,x^3,1)$, then $I^+(p) = M$ for every p in M). The existence of closed timelike curves in M is, from the point-of-view of physics, a most unfortunate state of affairs since it does violence to our most basic notions of causality. An observer whose worldline is closed might conceivably set off on a journey, return before his departure and decide not to go after all. While perhaps not beyond the realm of possibility such a situation at very least would leave physics in some disarray and we shall eventually formulate a condition ("stable causality") which, when imposed upon all of our models of the event world, will prohibit this and other such anomolous behavior. For the present we simply say that a spacetime M satisfies the *chronology condition* if $p \notin I^+(p)$ for each $p \in M$ and regard an M which does not satisfy this condition as physically unrealistic.

Our final objective in this section is to record our first indication of the influence of topology on physics by proving that any *compact* spacetime must fail the chronology condition (and so, in our view, be unworthy of serious consideration as a model of the event world). The proof of this, and a great many other results in our subject, depends upon the following crucial fact.

Lemma 3.3.10. For each p in M, $I^+(p)$ and $I^-(p)$ are open subsets of M.

Proof: Let q be in $I^+(p)$ (for $I^-(p)$ simply reverse the time orientation). We must show that there is a small open neighborhood of q in M contained entirely in $I^+(p)$. By assumption, there is a trip γ from p to q. The idea of the proof is as follows: Near each of its points M is "essentially identical to" its tangent space at that point. Choose a point r on the image of γ which is "sufficiently close" to q (in a geodesically convex normal neighborhood of q). Since the interior of the upper null cone in Minkowski space is open we can find a small neighborhood of q (thought of as lying in $T_r(M)$ via $\exp_r^{-1}$) contained in the future of r and therefore, by transitivity, of p as well. Here are the details:

Figure 3.8

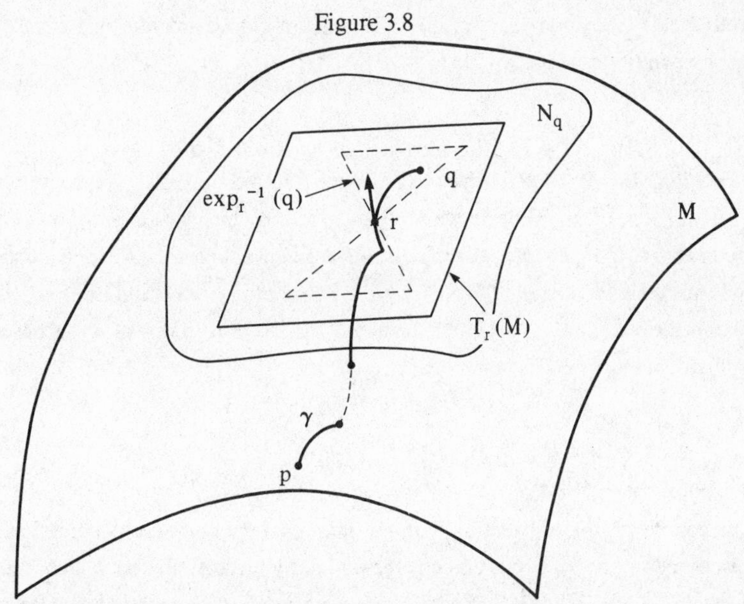

Let N_q be a geodesically convex normal neighborhood of q in M and select a point r other than q on the image of the terminal segment of γ and in N_q. By inserting one additional segment if necessary we can assume that the terminal segment of γ begins at r and therefore must be contained in N_q (since N_q is the image of a star-shaped neighborhood of 0 in $T_q(M)$ under $\exp_q$). Since N_q is geodesically convex it is the diffeomorphic image of some star-shaped open set U in $T_r(M)$ under $\exp_r$. Thus $\exp_r^{-1}(q)$ exists and is that tangent vector v in $T_r(M)$ with the property that the unique geodesic in M that fits v at r reaches q in parameter distance one. This geodesic segment is contained in N_q. As a point set it must coincide with the terminal segment of γ (again by geodesic convexity). Since q is the future endpoint of γ, both of these geodesics reach q in positive parameter distance so their respective parameters are related by an equation of the form $t = ms + b$ with $m > 0$ (Theorem 3.3.6). Their tangent vectors are therefore parallel and have the same direction. Since the terminal segment of γ is timelike and future directed, $\exp_r^{-1}(q)$ is likewise timelike and future directed. Consequently, $\exp_r^{-1}(q)$ is in the interior V of the upper null cone at r in $T_r(M)$. Since $\exp_r$ is a homeomorphism on U, $\exp_r(U \cap V)$ is open in N_q and therefore also in M. Furthermore, since the image of any point in V under $\exp_r$ is on a future directed timelike geodesic from r, $\exp_r(U \cap V)$ is contained in $I^+(r)$ and therefore, by transitivity of $\ll$, in $I^+(p)$. Thus, $\exp_r(U \cap V)$ is the required open neighborhood of q contained in $I^+(p)$. Q.E.D.

Theorem 3.3.11. Any compact spacetime M contains closed timelike curves, i.e., fails the chronology condition.

Proof: $\{I^+(p): p \in M\}$ is an open cover for M since any element p of M is in the future of something in M (let v be any future-directed timelike tangent vector at $p, \mu = \mu(t), -\varepsilon < t < \varepsilon$, a geodesic which fits v at p and consider a point $\mu(t_0)$ on μ with $t_0 < 0$). Let $\{I^+(p_1), I^+(p_2), ..., I^+(p_n)\}$ be a finite subcover. We may assume that $I^+(p_1)$ is not contained in any $I^+(p_j)$ for $j \geq 2$ for otherwise we could delete $I^+(p_1)$ from the subcover. But then $p_1 \in I^+(p_j)$ for any $j \geq 2$ by transitivity of $\ll$. Consequently, $p_1 \in I^+(p_1)$ and the proof is complete. Q.E.D.

3.4. Stable Causality

We concluded the last section with the failure of the chronology condition in compact spacetimes. Any such spacetime is then, in our view, causally badly behaved and we shall grant it no credibility as a model of the event world. There are, however, noncompact spacetimes which also contain closed timelike curves (see Chapter 5 of [HE]) so that restricting our attention to noncompact manifolds will not solve all of our problems. Indeed, the existence of closed timelike curves itself is not the only way in which a spacetime can be causally misbehaved. Closed null curves can exist even when the chronology condition is satisfied (e.g., in the M constructed from that portion of $\mathcal{M}$ between $x^4 = 0$ and $x^4 = 1$ by "identifying" points represented by $(x^1, x^2, x^3, 0)$ and $(x^1 + 1, x^2, x^3, 1)$). It is possible for M to possess neither closed timelike nor closed null curves, but nevertheless to have timelike curves which continually enter, leave and re-enter arbitrarily small neighborhoods of some point. Many such possibilities exist and we would like to impose some condition on our spacetimes which will prohibit all of them. Fortunately, the insight of Stephen Hawking has provided us with just such a condition and one which is, moreover, physically quite natural. We shall formulate a definition which, although most convenient for our purposes, is rather far removed from physical intuition and then describe an equivalent formulation which we hope to convince the reader should be satisfied by an "reasonable" model of the event world.

We shall say that a spacetime M is *stably causal* if it admits a *global time function*, i.e., if there exists a smooth, real-valued function $T: M \to \mathbb{R}$ whose gradient ∇T is everywhere timelike $(g(\nabla T(p), \nabla T(p)) < 0$ for each p).

Exercise 3.4.1. Show that on $\mathcal{M}$ and $\mathcal{E}$ the fourth coordinate function for the standard coordinate patch is a global time function. Find one for $\mathcal{D}$.

Observe that stable causality obviously implies time orientability. Also, if $\alpha = \alpha(t)$ is any smooth timelike curve in M, then, by Lemma 3.3.2, the derivative of T along α is $\alpha'(t)[T] = g(\nabla T, \alpha'(t))$ and this is nonzero since ∇T and α' are both timelike (Corollary 1.3.2). Thus, T is monotone along any timelike curve. Moreover, by choosing the time orientation appropriately ($V = -\nabla T$) we can assume that *T increases along future-directed timelike curves.* In particular, of course, M can contain no closed timelike curves.

At first glance the existence of a global time function on M is intuitively very appealing. One has visions of a "cosmic" notion of time and perhaps a relativistic version of the long since abandoned "absolute time" of Newton. In fact, however, the physical interpretation of these global time functions is not so clear. In particular, care must be taken not to confuse such a T with the timelike coordinate function of some global coordinate patch. No such global coordinate system need exist in a given stably causal spacetime (again, $\mathcal{M}$ and $\mathcal{E}$ are quite special). The real physical significance of stable causality comes to light in a remarkable theorem of Hawking which we pause now to discuss briefly (we return to the interpretation of T in the next section).

In addition to the causality violations we have already mentioned Hawking observed that certain spacetimes which do not have closed timelike curves are nevertheless "on the verge" of having them in the sense that an arbitrarily small perturbation of the metric can produce a new spacetime which fails the chronology condition. Now, from the point-of-view of physics, the metric is an object constructed by making measurements (of space and time intervals) and no measurement is infinitely accurate. Indeed, quantum mechanics imposes a positive lower bound on the uncertainty of any such measurement. We do not, and cannot, know the metric of the event world with absolute certainty. It follows that any physically meaningful assumption about a spacetime must be insensitive to small perturbations of the metric. Thus, it seems that assumptions such as the chronology condition are inherently weak; more appropriate would be the stipulation that there are no closed timelike curves in any Lorentz metric for M which is "close" to g. This is Hawking's original notion of stable causality, with "close" defined in terms of an appropriate topology on the set of all Lorentz metrics for M. We prefer the approach taken in [SW2]: Let M be a spacetime with metric g (since we shall be dealing with different metrics on the same manifold M we shall temporarily employ such expressions as "the spacetime (M,g)"). Another Lorentz metric h on M is said to be *wider* than g if, for every nonzero tangent vector v with $g(\mathrm{v},\mathrm{v}) \le 0$, it is also true that $h(\mathrm{v},\mathrm{v}) < 0$, i.e., if h-null cones are "opened out" relative to g-null cones:

Figure 3.9

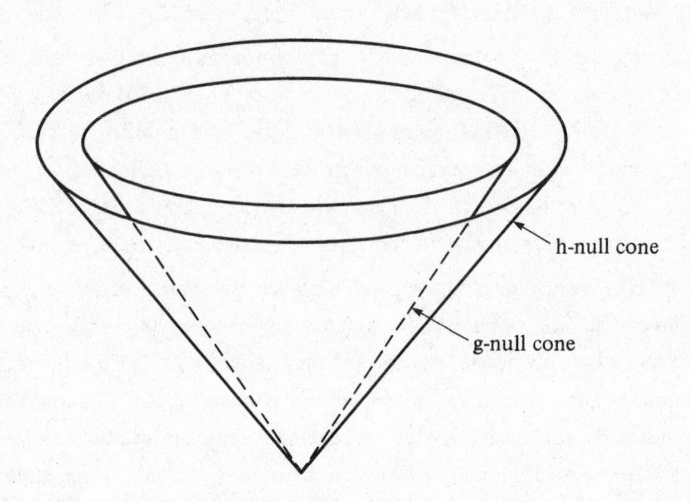

Observe that h has more timelike vectors than g so that the likelihood of finding closed timelike curves in (M,h) is greater than that of finding them in (M,g). More precisely, if (M,h) satisfies the chronology condition, then so does (M,g), but the converse is false. Indeed, if (M,h) satisfies the chronology condition, then (M,g) cannot even be "on the verge" of having closed timelike curves since g can be "perturbed" ("widened") slightly without producing such curves.

Exercise 3.4.2. Let (M,g) be a (time orientable) spacetime and V a smooth timelike vector field on (M,g). At each point p of M define a real-valued function h on $T_p(M) \times T_p(M)$ by (cartesian product)

$$h(\mathrm{v},w) = g(\mathrm{v},w) - g(\mathrm{v},V(p))g((w,V(p)) \tag{19}$$

for all v and w in $T_p(M)$. Show that h is a Lorentz metric on M and is, in fact, wider than g.

Hawking's theorem now takes the following form:

Theorem 3.4.1. A spacetime (M,g) is stably causal if and only if there exists a Lorentz metric h on M such that h is wider than g and (M,h) satisfies the chronology condition.

One direction is easy:

Exercise 3.4.3. Let $T: M \to \mathbb{R}$ be a global time function of (M,g) and take $V = -\nabla T$ in the construction in Exercise 3.4.2. Show that T is also a global time function for (M,h), i.e., that the h-gradient of T is h-timelike, and conclude that (M,h) satisfies the chronology condition.

The reverse implication is quite a bit more difficult; the argument is sketched on pages 199-201 of [HE]. We shall take Theorem 3.4.1 as sufficient motivation for assuming henceforth that *all of our spacetimes are stably causal.*

3.5. Global Hyperbolicity

The assumptions we have made of our spacetimes thus far (time orientability and stable causality) have, from the physical point-of-view been rather easy to live with. The next condition we propose is quite a bit stronger and not so easy to justify physically. Historically, Hawking's first general singularity theorem included this condition as one of its hypotheses, but subsequent refinements of this theorem substituted less controversial assumptions. We shall be content with the earliest version and direct those interested in the much more subtle variants to [HE], [O2] or [Pen] for detailed treatments.

We begin by considering an arbitrary global time function T on M (if there is one, there will be many). The level set $T^{-1}(r)$, $r \in \mathbb{R}$, have a number of properties which will interest us. Some are obvious. By continuity each such set is *closed.* Since T increases along future-directed timelike curves, no two points in $T^{-1}(r)$ can be "chronologically related", i.e., given p and q in $T^{-1}(r)$ neither $p \ll q$ nor $q \ll p$ can be true; we say that $T^{-1}(r)$ is *achronal.* It follows from our next theorem that any nonempty $T^{-1}(r)$ is a *3-dimensional smooth submanifold of M.*

Theorem 3.5.1. Let M be a smooth k-manifold with metric g and let $f: M \to \mathbb{R}$ be a smooth real-valued function on M. Then $S = \{p \in f^{-1}(r)\colon \nabla f(p) \text{ is nonzero}\}$ is either empty or a smooth $(k-1)$-dimensional submanifold of M.

Proof: Assume $S \neq \varnothing$ and fix a point p in S. We show that p has an open neighborhood in S that is diffeomorphic to an open set in $\mathbb{R}^{k-1}$. Choose an orthonormal basis $\{e_1, \ldots, e_k\}$ for $T_p(M)$ and an open, star-shaped neighborhood N of 0 in $T_p(M)$ which $\exp_p$ maps diffeomorphically onto the geodesically convex normal neighborhood N_p of p in M. Finally, let the coordinate patch $\exp_p$ establish a normal coordinate system $u^1, \ldots, u^k$ at p. Then $g_{ij}(p) = \eta_{ij} = \eta^{ij} = g^{ij}(p)$ so that $\nabla f\,(p) = \eta^{ij}\,\dfrac{\partial f}{\partial u^i}\,(p)\,e_j$. Since $\nabla f\,(p)$ is nonzero, some $\dfrac{\partial f}{\partial u^i}\,(p)$ is nonzero. Assume without loss of generality that $\dfrac{\partial f}{\partial u^k}\,(p) \neq 0$. Now consider the map $F: N \rightarrow \mathbb{R}^k$ defined by

$$F(u^1,\ldots,u^{k-1},u^k) = (u^1,\ldots,u^{k-1}, f(u^1,\ldots,u^{k-1},u^k)) \ .$$

F is obviously smooth and the Jacobian of F at $(0,\ldots,0)$ in $T_p(M) \sim \mathbb{R}^k$ is $\dfrac{\partial f}{\partial u^k}(p)$ which is nonzero. According to the Inverse Function Theorem (Theorem 3.2.1) there exists an open set U in N such that $F \mid U$ is a diffeomorphism onto some open neighborhood of $(0,\ldots,0,f(p))$ in $\mathbb{R}^k$. Thus, $\exp_p \circ (F \mid U)^{-1}: F(U) \rightarrow M$ is a coordinate patch at p in M whose image is $\exp_p(U)$ (an open neighborhood of p in M). Thus, $\exp_p(U) \cap S$ is an open neighborhood of p in S which consists precisely of those points in $\exp_p(U)$ whose kth coordinate relative to $\exp_p \circ (F \mid U)^{-1}$, i.e., $f(u^1,\ldots,u^k)$, is fixed at r.

Exercise 3.5.1. Complete the argument. Q.E.D.

Lemma 3.5.2. If M is a k-dimensional manifold with metric and $f: M \rightarrow \mathbb{R}$ is smooth, then ∇f is normal to the level hypersurfaces of f, i.e., if $\mathrm{v} \in T_p(S)$, where $S = \{p \in f^{-1}(r): \nabla f(p) \text{ is nonzero}\}$, then $<\nabla f(p), \mathrm{v}> = 0$.

Proof: Since $\mathrm{v} \in T_p(S)$ implies $\mathrm{v} \in T_p(M)$ we may use Lemma 3.3.2 to compute $<\nabla f(p), \mathrm{v}> = \mathrm{v}[f] = \dfrac{d}{dt} f(\alpha(t))\Big|_{t=0}$, where α is any curve which fits v at p. But, by assumption, there is curve α *in S* which fits v at p and f is constant along any curve in S so $\dfrac{d}{dt} f(\alpha(t))\Big|_{t=0} = 0$.

Q.E.D.

Now let us return to our global time function $T: M \rightarrow \mathbb{R}$ on the spacetime M and consider a nonempty level set $S = T^{-1}(r)$. Since ∇T is everywhere timelike it is, is particular, never zero so Theorem 3.5.1 guarantees that S is a smooth 3-dimensional submanifold of M. Moreover, Lemma 3.5.2 and Corollary 1.3.2 imply that each $T_p(S)$ must consist entirely of spacelike vectors (and zero). A 3-dimensional submanifold S of a spacetime M every tangent space of which consists of zero and spacelike vectors in M is called a *spacelike submanifold* of M. To summarize then we have found that any nonempty level set for a global time function must be a closed, achronal, spacelike, 3-dimensional smooth submanifold of M. Despite the difficulties one encounters in attempting to supply global time functions with a direct physical significance these properties are essentially the closest one can come in an abstract setting to the intuitive notion of an "instantaneous 3-

space" (all space at some "instant"). For this reason we shall hencefore refer to any nonempty level set for a global time function on M as a *spacelike slice* of M. The most obvious examples, of course, are the u^4 = constant hypersurfaces in $\mathcal{M}$, $\mathcal{D}$ and $\mathcal{E}$ (diffeomorphic to $\mathbb{R}^3$ in $\mathcal{M}$ and $\mathcal{E}$ and S^3 in $\mathcal{D}$). Two more useful examples are obtained as follows: Let $N^+ = \{(u^1,...,u^4) \in \mathcal{M} : u^4 > ((u^1)^2 + (u^2)^2 + (u^3)^2)^{1/2}\}$ (the interior of the upper null cone in $\mathcal{M}$, regarded as a spacetime in its own right). Define $T : N^+ \to \mathbb{R}$ by $T(u^1,...,u^4) = ((u^4)^2 - (u^1)^2 - (u^2)^2 - (u^3)^2)^{1/2}$. Then T is smooth on N^+ and $\nabla T = -\frac{u^1}{T} e_1 - \frac{u^2}{T} e_2 - \frac{u^3}{T} e_3 + \frac{u^4}{T} e_4$ so $<\nabla T, \nabla T> =$ $= \frac{(u^1)^2 + (u^2)^2 + (u^3)^2 - (u^4)^2}{T^2}$ which is negative on N^+. Thus, T is a global time function on N^+. A typical spacelike slice for T is the upper branch ($u^4 > 0$) of the hyperboloid

$$(u^1)^2 + (u^2)^2 + (u^3)^2 - (u^4)^2 = -1 \ .$$

The same function T defined on the interior of the lower null cone in $\mathcal{M}(N^- = \{(u^1,...,u^4) \in \mathcal{M} : u^4 < -((u^1)^2 + (u^2)^2 + (u^3)^2)^{1/2})\}$ is a global time function whose spacelike slices are the lower branches ($u^4 < 0$) of these same hyperboloids.

Thinking of a spacelike slice S intuitively as "all space at some instant" we now wish to address the question of S's "domain of influence", i.e., how much of M is in some sense "determined" by data on S. The definition we propose suggests that what happens at an event p in M is "determined" on S if every past endless smooth timelike curve from p meets S so that any effect which could influence p is "registered" on S. Again we prefer to phrase the definition in terms of trips: Let S be a spacelike slice in M. The *future Cauchy development* of S, denoted $D^+(S)$, is the set of all $p \in M$ such that every past endless trip (timelike curve) from p meets S. The *past Cauchy development* of S, denoted $D^-(S)$, is defined by replacing "past" with "future". The *Cauchy development* (*domain of dependence*) of S is

$$D(S) = D^+(S) \cup D^-(S) \ .$$

In $\mathcal{M}$ each $u^4 = u_0^4$ (constant) is a spacelike slice S with $D^+(S) =$ $= \{(u^1,...,u^4) \in \mathcal{M} : u^4 \geq u_0^4\}$ and $D(S) = \mathcal{M}$. Analogous statements are true in $\mathcal{D}$ and $\mathcal{E}$. In N^+, the slice $S = \{(u^1,...,u^4)$: $(u^1)^2 + (u^2)^2 + (u^3)^2 - (u^4)^2 = -1$, $u^4 > 0\}$ has $D^+(S)$ equal to S and everything "above":

Figure 3.10

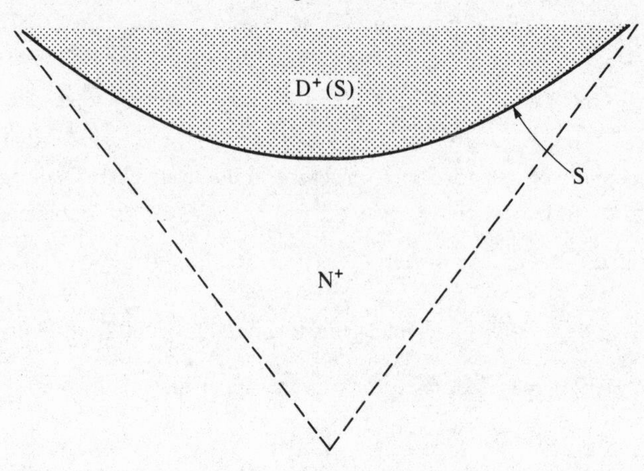

On the other hand, in N^- the slice $S = \{(u^1,...,u^4)\colon (u^1)^2 + (u^2)^2 + (u^3)^2 - (u^4)^2 = -1,\ u^4 < 0\}$ has $D^+(S)$ as shown in Figure 3.11.

Figure 3.11

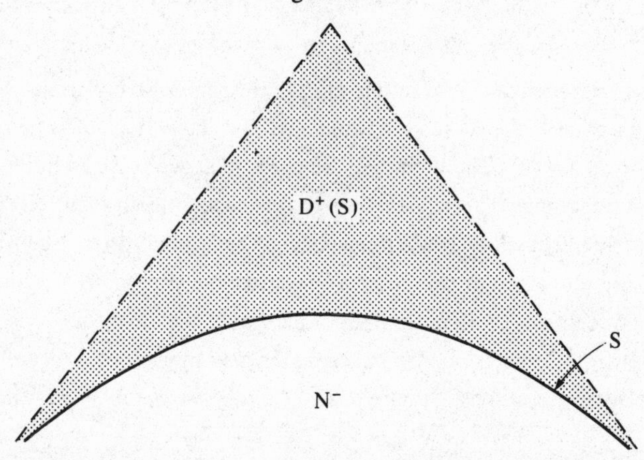

Our last example is instructive in that it indicates the effect on $D^+(S)$ of cutting a "hole" in our spacetime. We let $M = \mathcal{M} - \{(0,0,0,0)\}$ and $S = \{(u^1,...,u^4)\colon u^4 = 0\}$. Then $D^+(S)$ is as shown in Figure 3.12.

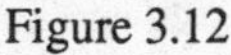

Figure 3.12

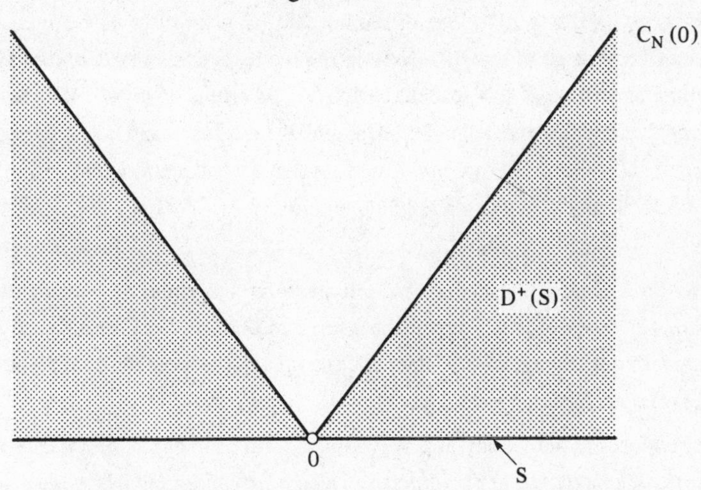

The "shadow" in $D^+(S)$ corresponds physically to the fact that information could, at least in principle, be emitted from the hole, influencing the physics in the shadow region without that information begin registered on S.

Exercise 3.5.2. Let S be a spacelike slice in the spacetime M. Show that (a) $S \subseteq D^+(S)$, and (b) $D^+(S)$ is closed in M. **Hint for** (b): Let $p \in D^+(S)$ and consider $N_p \cap I^+(q)$, where N_p is a geodesically convex normal neighborhood of p which misses S and q is point in N_p to the past of p on some past endless trip that misses S.

Remarks. The physical interpretation of $D^+(S)$ is somewhat tenuous since we have intentionally ignored the fact that causal effects can be propagated along null as well as timelike curves. Allowing such curves in the definition complicates the mathematics somewhat and gives a *smaller* $D^+(S)$ whose closure in M is our $D^+(S)$. This is the approach taken in [HE] which may be consulted for more details.

From the causal point-of-view one would be most interested in a spacelike slice $S = T^{-1}(r)$ for which $D(S) = M$ (intuitively, anything that happened "before" $T = r$ is recorded on S and determines everything that happens "after" $T = r$). Such slices exist in some spacetimes (e.g., u^4 = constant in $\mathcal{M}, \mathcal{D}$ and $\mathcal{E}$), but not in others (e.g., $\mathcal{M} - \{(0,0,0,0)\}$). Traditionally, a "Cauchy surface" in a spacetime M is defined to be a closed, achronal, spacelike, 3-dimensional submanifold S of M for which $D(S) = M$. M is then said to be "globally hyperbolic" if it contains such a Cauchy surface. One can then prove that any such M must be stably causal and, indeed, must have a global time function T *all* of whose level sets $T^{-1}(r)$ are diffeomorphic "Cauchy surfaces". We prefer to evade the issue of proving all of this by incorporating the results into our definitions: A *Cauchy surface* in a spacetime M is a spacelike slice S in M with $D(S) = M$. M is said to be *globally hyperbolic* if there is defined on M a global time function T all of whose nonempty level sets $T^{-1}(r)$ are diffeomorphic Cauchy surfaces (such a T is called a *Cauchy time function*). $\mathcal{M}$, $\mathcal{D}$ and $\mathcal{E}$ are all globally hyperbolic.

Exercise 3.5.3. Show that any globally hyperbolic spacetime M is diffeomorphic to a product $S \times \mathbb{R}$, where S is a 3-dimensional manifold. *Hint.* Let T be a Cauchy time function. Show that the image of T is an open interval and compose with a diffeomorphism of that interval onto $\mathbb{R}$ (Exercise 3.2.4).

We regard global hyperbolicity as a very strong form of the principle of determinism and *henceforth assume that this condition is satisfied by all of our models of the even world.*

3.6. Curvature and the Einstein Field Equations

Let $D = \{(u^1, u^2, u^3, u^4) \in \mathbb{R}^4 : -\infty < u^1, u^4 < \infty, 0 < u^2 < \pi, -\pi < u^3 < \pi\}$ and suppose we are told that a certain map χ on D is a coordinate patch for some spacetime M relative to which the nonzero metric components are given by

$$g_{22} = \frac{1}{4}(u^1 - u^4)^2,\ g_{33} = \frac{1}{4}(u^1 - u^4)^2 \sin^2 u^2,\ g_{14} = g_{41} = -\frac{1}{2} .$$

Observe that these metric components change from point to point in M. In general, nonconstancy of metric components can be traced to one of two possible sources. Either the geometry of the event world is really changing from point to point because M models a nontrivial gravitational field (this is the case, for example, in $\mathcal{D}$ and $\mathcal{E}$), or we are being deceived into believing that such is the case by an unfortunate choice of coordinates (see Exercise 3.2.17 or, better yet, think about the range of possibilities Exercise 3.2.16 presents for the coordinate representation of the metric on Minkowski spacetime). How are we to know? In this particular case at least the answer is not difficult.

Exercise 3.6.1 Define a new coordinate system for the spacetime M above by setting $u^1 = \bar{u}^1 + \bar{u}^4$, $u^2 = \bar{u}^2$, $u^3 = \bar{u}^3$ and $u^4 = \bar{u}^4 - \bar{u}^1$. Use (4) of Exercise 3.2.16 to calculate the metric components in the barred coordinate system and observe that these are precisely the metric components for Minkowski spacetime relative to spherical spatial coordinates (Exercise 3.2.17).

We conclude then that M is just a disguised version of $\mathcal{M}$ and no gravitational field is involved. In general, finding a coordinate transformation which carries given metric components to more familiar ones is a highly nontrivial task requiring considerable ingenuity. On the other hand, there is a more or less routine (albeit quite tedious) calculation one can perform to decide whether or not a coordinate transformation exists which converts a given set of metric components into the standard metric components on $\mathcal{M}$, i.e., whether or not a given M is locally "flat" (indistinguishable from $\mathcal{M}$). In order to trace the origins of this "test" for local flatness we shall sketch an argument that is carried out in detail on pages 183-188 and 200-204 of Volume III of [Sp2].

Begin with an arbitrary coordinate patch $\chi: D \to M$ on the spacetime M (obvious alterations in the argument yield the same result of any Riemannian or Lorentzian manifold). Let g_{ij} be the metric components relative to χ. We ask whether or not there exists another coordinate patch $\bar{\chi}: \bar{D} \to M$ with $\chi(D) \cap \bar{\chi}(\bar{D}) \neq \varnothing$ and relative to which the metric components $\bar{g}_{ij}$ are given by $\bar{g}_{ij} = \eta_{ij}$ *throughout* $\chi(D) \cap \bar{\chi}(\bar{D})$ (observe that, at each fixed point p of M, this condition is automatically satisfies *at* p by any set of Minkowski normal coordinates at p). If such a coordinate patch exists, then the coordinate transformation $\bar{u}^i = \bar{u}^i(u^1, u^2, u^3, u^4)$, $i = 1,2,3,4$, must, by Exercise 3.2.16 satisfy

$$g_{ij} = \frac{\partial \bar{u}^\alpha}{\partial u^i} \frac{\partial \bar{u}^\beta}{\partial u^j} \eta_{\alpha\beta}, \quad i,j = 1,2,3,4, \tag{20}$$

throughout $\chi(D) \cap \bar{\chi}(\bar{D})$. Rewrite (20) as

$$g_{ij} = \frac{\partial \bar{u}^\alpha}{\partial u^i} \frac{\partial \bar{u}^\alpha}{\partial u^j} \varepsilon_\alpha, \quad i,j = 1,2,3,4 \tag{21}$$

where

$$\varepsilon_\alpha = \begin{cases} 1 & \text{if } \alpha = 1,2,3 \\ -1 & \text{if } \alpha = 4\,. \end{cases}$$

Observe that, since the g_{ij} are known, (21) can be regarded as a set of partial differential equations for the sought after coordinate transformation $\bar{u}^\alpha = \bar{u}^\alpha(u^1, u^2, u^3, u^4)$, $\alpha = 1,2,3,4$. Differentiation of (20) with respect to u^k and a few algebraic maneouvers give

$$\frac{1}{2}\left[\frac{\partial g_{ij}}{\partial u^k} + \frac{\partial g_{ik}}{\partial u^j} - \frac{\partial g_{jk}}{\partial u^i}\right] = \frac{\partial^2 \bar{u}^\alpha}{\partial u^j \partial u^k} \frac{\partial \bar{u}^\alpha}{\partial u^i} \varepsilon_\alpha, \quad i,j,k = 1,2,3,4. \tag{22}$$

Multiplying (22) by $g^{i\gamma}\dfrac{\partial\bar{u}^{\lambda}}{\partial u^{\gamma}}$ and summing as indicated yields

$$\frac{\partial}{\partial u^k}\left(\frac{\partial\bar{u}^{\lambda}}{\partial u^j}\right)=\Gamma^{\alpha}_{jk}\,\frac{\partial\bar{u}^{\lambda}}{\partial u^{\alpha}}\;,\lambda,j,k=1,2,3,4, \tag{23}$$

which is a system of partial differential equations for the $\bar{u}^{\lambda}$ to which standard existence theorems apply. Indeed, the "integrability conditions" which must be satisfied in order that (23) admit a smooth solution $\bar{u}^{\lambda}=\bar{u}^{\lambda}(u^1,u^2,u^3,u^4)$, $\lambda=1,2,3,4$, are

$$\frac{\partial\Gamma^{\gamma}_{kj}}{\partial u^1}-\frac{\partial\Gamma^{\gamma}_{1j}}{\partial u^k}+(\Gamma^{\mu}_{kj}\,\Gamma^{\gamma}_{1\mu}-\Gamma^{\mu}_{1j}\,\Gamma^{\gamma}_{k\mu})=0,\gamma,j,k,1=1,2,3,4\;. \tag{24}$$

We find then that *necessary* conditions for the existence of our coordinate patch χ are that the functions on the left-hand side of (24) vanish throughout the image of χ. That these are also *sufficient* conditions is not obvious, but is true (see pages 200-204, Volume III of [Sp2]).

It goes without saying that calculating the $4^4=256$ functions on the left-hand side of (24) from the metric components g_{ij} and the definition of the Christoffel symbols is an arduous task (although not quite as bad as it might seem since there are actually only 20 independent functions). Nevertheless, these functions reappear again and again in differential geometry and relativity whenever one must contend with questions of "curvature" versus "flatness". They were first singled out for consideration by Riemann in his generalization to dimension n of Gauss' differential geometry of surfaces in $\mathbb{R}^3$ and are named in his honor. If M is any k-dimensional smooth manifold with metric g and if $\chi\colon D\to M$ is a coordinate patch for M, then we define the *components relative to* χ *of the Riemann curvature tensor* R for M by

$$R^a_{\ bcd}==\frac{\partial\Gamma^a_{bd}}{\partial u^c}-\frac{\partial\Gamma^a_{bc}}{\partial u^d}+(\Gamma^{\alpha}_{bd}\,\Gamma^a_{\alpha c}-\Gamma^{\alpha}_{bc}\,\Gamma^a_{\alpha d}),\;a,b,c,d=1,\ldots,k\;. \tag{25}$$

Remarks. We have not defined the unmodified word "tensor" and will have no need to do so. However, our experience with "4-tensors" in Chapter 2 should leave little room for doubt as to the proper definition. Recall that a 4-tensor of contravariant rank 1 and covariant rank 3 is an "object" which is described in each admissible frame of reference by $4^4=256$ numbers $T^a_{\ bcd}$, $a,b,c,d=1,2,3,4$, with the property that if two admissible frames are related by the Lorentz transformation $\bar{x}^a=\Lambda^a_{\ b}x^b$, $a=1,2,3,4$, then the numbers which describe the 4-tensor in the two frames are related by

$$\bar{T}^a_{bcd} = \Lambda^a_\alpha \bar{\Lambda}^\beta_b \bar{\Lambda}^\gamma_c \bar{\Lambda}^\delta_d T^\alpha_{\beta\gamma\delta} \tag{26}$$

for $a,b,c,d = 1,2,3,4$. Alternatively, it is a real-valued, 4-linear operator on 4-vectors whose value at (U,V,W,X) is

$$T(U,V,W,X) = T^a_{bcd} U_a V^b W^c X^d \ .$$

Observing that $\Lambda^a_\alpha = \dfrac{\partial \bar{x}^a}{\partial x^\alpha}$, $\bar{\Lambda}^\beta_b = \dfrac{\partial x^\beta}{\partial \bar{x}^b}$, etc. one can write the transformation law (26) as

$$\bar{T}^a_{bcd} = \frac{\partial \bar{x}^a}{\partial x^\alpha} \frac{\partial x^\beta}{\partial \bar{x}^b} \frac{\partial x^\gamma}{\partial \bar{x}^c} \frac{\partial x^\delta}{\partial \bar{x}^d} T^\alpha_{\beta\gamma\delta} \ .$$

The transformation law for the Christoffel symbols given in Exercise 3.3.3 and a healthy supply of persistence show that if $\bar{\chi}$ is another coordinate patch for the k-manifold M which overlaps χ, then the quantities $\bar{R}^a_{bcd}$ defined by (25) with barred coordinates are related to the R^a_{bcd} by

$$\bar{R}^a_{bcd} = \frac{\partial \bar{u}^a}{\partial u^\alpha} \frac{\partial u^\beta}{\partial \bar{u}^b} \frac{\partial u^\gamma}{\partial \bar{u}^c} \frac{\partial u^\delta}{\partial \bar{u}^d} R^\alpha_{\beta\gamma\delta}$$

and it is this transformation law which entitles the Riemann curvature functions defined by (25) to the name "tensor".

Exercise 3.6.2. Define "contravariant" and "covariant" components of a tangent vector v in $T_p(M)$ relative to each coordinate patch χ at p and interpret the Riemann curvature tensor R as an operator on tangent vectors. *Hint.* Look at the corresponding definitions for 4-vectors in Chapter 2 and keep in mind that η_{ij} is the *metric* for $\mathcal{M}$.

Observe that the transformation law (27) makes it clear that if $R^\alpha_{\beta\gamma\delta} = 0$ for all $\alpha,\beta,\gamma,\delta = 1,\ldots,k$, in one coordinate system, then $\bar{R}^a_{bcd} = 0$ for all $a,b,c,d = 1,\ldots,k$, in any other coordinate system as well. Thus, "the Riemann curvature of M is zero" is an invariant statement, not effected by a change in coordinates. In particular, if one calculated the components of the curvature tensor for the spherical coordinate patch for $\mathcal{M}$ (Exercise 3.2.16) the result must be identically zero since it is obviously zero for the standard coordinate patch (where all Christoffel symbols are zero). Intuitively, the local "flatness" of a manifold is not a matter of perspective, i.e., of one's coordinate system.

We will have no occasion to deal with the detailed calculation of components for the Riemann tensor of specific manifolds so we shall content ourselves with quoting the results of these calculations for $\mathcal{D}$ and $\mathcal{E}$. If g_{ij} are the metric components for deSitter spacetime $\mathcal{D}$ relative to some coordinate patch (for example, that described in Exercise 3.2.18), then

$$R^a_{bcd} = \delta^a_d g_{bc} - \delta^a_c g_{bd} \quad (\text{in } \mathcal{D}) \tag{28}$$

(this is most often proved by showing that $\mathcal{D}$ has constant "sectional curvature" $K = 1$; see [O2]). The most efficient method of calculating the curvature tensor for Einstein-deSitter spacetime $\mathcal{E}$ is by way of the so-called "curvature forms" of Cartan (see pages 32-33 of [SW2]). The result is as follows: For $i, j = 1,2,3$,

$$\left.\begin{aligned} R^i_{jij} &= -R^i_{jji} = \frac{4}{9}(u^4)^{-2} \\ R^i_{4i4} &= -R^i_{44i} = R^4_{ii4} = -R^4_{i4i} = \frac{2}{9}(u^4)^{-2} \end{aligned}\right\} \quad (\text{in } \mathcal{E}) \tag{29}$$

and all remaining R^a_{bcd} are zero (here u^4 is the fourth coordinate of the point in $\mathcal{E}$ at which R is being calculated).

Our primary concern will be with three "tensors" derivable from the Riemann curvature tensor (we shall leave it to the reader to define the appropriate type of tensor and verify that the objects we describe are of these types). The *Ricci tensor* Ric has components in each coordinate patch that are obtained by "contracting" the corresponding components of R. Specifically, the components R_{ab} of Ric are defined in each coordinate system by

$$R_{ab} = R^\alpha_{a\alpha b} \ . \tag{30}$$

In $\mathcal{M}$, of course Ric is identically zero. In deSitter spacetime we calculate from (28) that $R_{ab} = \delta^\alpha_b \, g_{a\alpha} - \delta^\alpha_\alpha g_{ab} = g_{ab} - 4g_{ab}$, i.e.,

$$R_{ab} = -3g_{ab} \qquad (\text{in } \mathcal{D}) \ . \tag{31}$$

Exercise 3.6.3. Use (29) to show that the Ricci tensor for Einstein-deSitter spacetime has nonzero components relative to the standard coordinate patch given by

$$R_{aa} = \frac{2}{3}(u^4)^{-2}, \quad a = 1,2,3,4 \qquad (\text{in } \mathcal{E}) \ . \tag{32}$$

The *scalar curvature* S is obtained by "contracting" Ric, i.e.,

$$S = R^a_a = g^{ab} R_{ab} \ . \tag{33}$$

Thus, in $\mathcal{D}$, $S = g^{ab}(-3g_{ab}) = -3\delta^a_a = -3(4) = -12$, i.e., $\mathcal{D}$ has constant scalar curvature.

Exercise 3.6.4. Show that the scalar curvature in $\mathcal{E}$ is given by $S = \frac{4}{3}(u^4)^{-2}$.

Remarks. The scalar curvature of M is a real-valued function on M which is often regarded as a gross numerical measure of the extent to which M is "curved" at each point. $\mathcal{M}$ isn't curved anywhere, $\mathcal{D}$ has the same degree of curvature everywhere and $\mathcal{E}$'s curvature becomes unbounded as $u^4 \to 0$, i.e., as one recedes into the past. This last example is particularly interesting in that it indicates that our cosmological model becomes "singular" in the past (as one approaches the "big bang").

The *Einstein tensor* G of M is now defined by $G = \text{Ric} - \frac{1}{2}Sg$, i.e.,

$$G_{ab} = R_{ab} - Sg_{ab} \ .$$

For $\mathcal{M}, G$ is zero of course. For $\mathcal{D}, G$ is a multiple of the metric ($G = 3g$). For $\mathcal{E}$, $G_{44} = \frac{4}{3}(u^4)^{-2}$ and the remaining G_{ab} are zero. Our interest in the Einstein tensor stems from the the role it plays in Einstein's field equations, to which we now turn our attention.

Much of the early work in general relativity focused on the construction of specific spacetimes which modeled various gravitational fields of interest to physicists (that of the large scale distribution of galaxies in our universe, that of a static, spherically symmetric star, a rotating star, etc.). The constructions were carried out by solving a system of partial differential equations proposed by Einstein as the link between the matter and energy which give rise to the gravitational field (represented by the total energy-momentum tensor T of the system) and the metric g which models it. With the notation we have introduced these *Einstein field equations* take the form

$$G = 8\pi T \tag{34}$$

or, in somewhat more detail,

$$R_{ab} - \frac{1}{2}Sg_{ab} = 8\pi T_{ab}, \quad a,b = 1,2,3,4 \ . \tag{35}$$

Remark. Einstein (briefly) considered a variant of these equations which one often still encounters in the literature. Specifically, the *field equations with cosmological constant* have the form

$$G - \Lambda g = 8\pi T \ , \tag{36}$$

where Λ is a (generally positive and very small) constant. For the history and significance of Λ we refer to [Pa].

Observe that the left-hand side of (35) is a purely geometrical object, defined for any manifold with metric, while the right-hand side is fundamentally a physically measured quantity. In order to gain some sense of the enormous complexity of these equations we recommend the following:

Exercise 3.6.5. Write out the left-hand side of the Einstein equations (35) entirely in terms of the metric components g_{ab} and their derivatives.

The result of Exercise 3.6.5 is a system of nonlinear partial differential equations for the unknown metric components g_{ab}. The nonlinearity has its origin in the fact that, while T_{ab} contains a contribution from each of the relevant electromagnetic and matter fields, it does *not* contain a term which reflects the fact that the gravitational field itself contains energy which, being equivalent to an additional mass contribution, must "gravitate". In relativity a gravitational field "feeds upon itself" and this is reflected mathematically in the nonlinearity of the field equations.

The reader may have noticed another, rather disconcerting, aspect of the Einstein equations. All of the quantities which appear in (35), whether geometrical or physical, are expressed in terms of some coordinate system on the manifold M. What coordinate system? Indeed, what manifold? The equations seem to be expressed in terms of the very thing they are being used to construct and, in a sense, this is true. To solve (35) one must begin with a "guess" (based upon one's physical intuition concerning the field being modeled) as to what at least one coordinate patch on the sought after manifold might look like (e.g., spherical symmetry would suggest one guess, homogeneity and isotropy another, etc.). Moreover, such a guess is possible only if the field is assumed to possess exact symmetries which, of course, are never found in nature. We shall not pursue these matters here since our results will not require the full strength of the field equations. Indeed, the work of Hawking and Penrose on "global" results in relativity was motivated principally by the desire to know whether or not some rather remarkable properties of exact solutions to the field equations were a consequence only of physically unrealistic symmetry assumptions. Any standard text on general relativity will contain sections on solving the field equations; we particularly recommend [ABS]. A brief survey of many of the known exact solutions is available in [HE].

Remark. Strictly speaking any Lorentz metric g on a spacetime M can be regarded as a solution to the field equations. One need only construct from g the Einstein tensor G and then *define* the energy-momentum tensor by $T = \frac{1}{8\pi}G$. This is cheating, of course, since such a T will not in general correspond to any realistic distribution of mass/energy. Indeed, this is the case for deSitter spacetime $\mathcal{D}$ where $G = 3g$ so $T = (3/8\pi)g$ and, for any timelike V, $T_{ab}V^aV^b = (3/8\pi)g_{ab}V^aV^b = (3/8\pi)g(V,V) < 0$ and so corresponds to a *negative* energy density (alternatively, one can regard deSitter spacetime as the empty space, i.e., $T = 0$, solution to the equations (36) with cosmological constant $\Lambda = 3$).

The road which led Einstein from his earliest thoughts on a relativistic theory of gravitation to his field equations was extraordinarily long and difficult as certainly befits what has been called the greatest achievement of any single human mind. We will not presume to "summarize" the route. A marvelous glimpse into the history of the equations is available in [Pa], while a detailed synopsis of Einstein's "derivation" can be found, for example, in [ABS].

An alternate version of the field equations will be useful. Beginning with (35) we multiply by g^{ab}, sum as indicated and define the *trace* of T by trace $T = g^{ab}T_{ab}$, thus yielding

$$g^{ab}(R_{ab} - \frac{1}{2}Sg_{ab}) = 8\pi g^{ab}T_{ab}$$

$$S - \frac{1}{2}S(4) = 8\pi(\text{trace } T)$$

$$S = -8\pi(\text{trace } T)$$

so that (35) can be written

$$R_{ab} = 8\pi(T_{ab} - \frac{1}{2}(\text{trace } T)\, g_{ab}), \quad a,b = 1,2,3,4 \ . \tag{37}$$

Exercise 3.6.6. Show that the *empty space field equations* $G = 0$ are equivalent to $\text{Ric} = 0$. A solution to these equations is called either a *vacuum solution* or *Ricci flat*.

As in Chapter 2 we shall say that an energy-momentum tensor T satisfies the *strong energy condition* if, for every timelike vector V,

$$T_{ab}V^aV^b \geq \frac{1}{2}(\text{trace } T)\, g_{ab}\, V^b \ . \tag{38}$$

This is regarded as a physically reasonable requirement on the energy-momentum tensor and is, in fact, satisfied by all normal matter and energy fields (see [HE]). Combined with the field equations the strong energy condition implies that *for all timelike* V,

$$R_{ab}\, V^a V^b \geq 0 \ , \tag{39}$$

i.e.,

$$\text{Ric}\,(V,V) \geq 0 \ . \tag{40}$$

The requirement that (40) be satisfied for all timelike V is called the *timelike convergence condition* and is all that we shall retain of the field equations. It is one of the hypotheses of Hawking's singularity theorem and is often conceived of as a mathematical statement of the assumption that gravity is always attractive (see page 122-123 of [SW2]).

Exercise 3.6.7. An energy-momentum tensor T is said to satisfy the *weak energy condition* if $T_{ab}V^aV^b \geq 0$ for all timelike V. Show that if this condition is satisfied, then $T_{ab}W^aW^b \geq 0$ for all null W. Combined with the field equations this then implies the *null convergence condition*

$$\text{Ric}\ (W,W) \geq 0 \tag{41}$$

for all null W.

3.7. Mean Curvature and the Expansion of the Universe

In its early years general relativity had many great successes. The unaccountable excess in the advance of the perihelion of Mercury observed by LeVerrier sixty years earlier was handled immediately and with remarkable accuracy (see [ABS]). Eddington's dramatic confirmation of the bending of light in a gravitational field precipitated what today could only be called a "media event" and overnight transformed Einstein from a physicist into a legend (see [Pa] for the history and [ABS] for the mathematical details). But perhaps the most awe inspiring was one which, in a sense, never happened. We have already observed (see the Remark on page 100) that in our comological model $\mathcal{E}$ the galaxies of our universe are receding from one another and this is by no means a peculiarity of $\mathcal{E}$ alone. Einstein was the first to apply general relativity to the study of the large scale structure of the universe and was well aware of the dynamic, evolutionary nature of the cosmological solutions to his field equations $\text{Ric} - \frac{1}{2}Sg = 8\pi T$. Now, one must understant that at this time (1915-1917) such a notion was absolutely beyond comprehension.

Throughout all of human history the universe was regarded as fixed and immutable and the idea that it might actually be changing was inconceivable. And so, for perhaps the only time in his life, Einstein lost faith. Despite his profound belief in the validity of his field equations he simply could not reconcile himself with their prediction of a non-static universe and so he introduced, albeit very reluctantly, the modified field equations $\mathrm{Ric} - \frac{1}{2} Sg - \Lambda g = 8\pi T$ which do admit a static cosmological solution. Thirteen years later Hubble discovered that the universe is, in fact, expanding and Einstein repented, withdrawing the cosmological constant Λ and calling its introduction "the greatest blunder of my life". Einstein was thus denied what would certainly have been the greatest achievement of his (and arguably any other) scientific theory: predicting, before any evidence was available, the expansion of the universe.

Nevertheless, the universe *is* expanding and this expansion is one of the hypotheses of Hawking's theorem. To get at a precise mathematical statement of this hypothesis we will require a generalization of the familiar notion of "divergence" from vector calculus. Thus, we consider a smooth vector field W on the manifold M with metric. If $\alpha: I \to M$ is any smooth curve in M we define the *restriction of W to α* to be the vector field along α denoted $W \mid \alpha$ and defined by

$$(W \mid \alpha)\,(t) = W(\alpha(t))$$

for each t in I.

Remark. In a sense this process can be "reversed" also:

Exercise 3.7.1. Let $\alpha: I \to M$ be a smooth curve in the manifold M, V a smooth vector field along α and $t_0 \in I$. Show that there exists an open neighborhood of $\alpha(t_0)$ in M and a smooth vector field W on this open set such that $(W \mid \alpha)\,(t) = V(t)$ for all t in some interval about t_0.

Now let p be a point of M and $\mathrm{v} \in T_p(M)$. We define the *covariant derivative* of W *at p in the direction* v, denoted $D_{\mathrm{v}} W$, by selecting a smooth curve α which fits v at p and setting

$$D_{\mathrm{v}}\, W = D_{\alpha'}(W \mid \alpha) \tag{42}$$

(see (14) of section 3.3). Thus, in any coordinate patch,

$$D_{\mathrm{v}} W = \left[\frac{dW}{dt} + \Gamma^r_{ij}\, W^i \frac{du^j}{dt} \right] \chi_r\ , \tag{43}$$

where $\alpha(t) = \chi(u^1(t),...,u^k(t))$. Observe that, at each p,

$$D_{\mathrm{v}} W(p) = \left[\frac{\partial W^r}{\partial u^j}(p) \frac{du^j}{dt}(0) + \Gamma^r_{ij} W^i(p) \frac{du^j}{dt}(0) \right] \chi_r(p)$$

so that

$$D_{\mathrm{v}} W = \left[\frac{\partial W^r}{\partial u^j} + \Gamma^r_{ij} W^i \right] \mathrm{v}^j \chi_r \tag{44}$$

and therefore the definition is independent of α. If V and W are both smooth vector fields we define the vector field $D_V W$ by

$$D_V W(p) = D_{V(p)} W \ .$$

Exercise 3.7.2. Prove each of the following:

1. $D_{\mathrm{v}} W$ is linear in W, i.e., $D_{\mathrm{v}}(c_1 W_1 + c_2 W_2) = c_1 D_{\mathrm{v}} W_1 + c_2 D_{\mathrm{v}} W_2$ for any real numbers c_1 and c_2.

2. If f is any smooth real-valued function, then $D_{\mathrm{v}}(fW) = \mathrm{v}[f]W(p) + f(p) D_{\mathrm{v}} W$.

The quantity in brackets in (44) is called the *covariant derivative of* W and is denoted $W^r_{;j}$, i.e.,

$$W^r_{;j} = \frac{\partial W^r}{\partial u^j} + \Gamma^r_{ij} W^i \tag{45}$$

in each coordinate system.

Exercise 3.7.3. Show that the quantities $\overline{W}^r_{;j}$ defined by (45) with barred coordinates are related to the $W^r_{;j}$ by

$$\overline{W}^r_{;j} = \frac{\partial \bar{u}^r}{\partial u^\alpha} \frac{\partial u^\beta}{\partial \bar{u}^j} W^\alpha_{;\beta}$$

so that the covariant derivative of W has the transformation law of what one would call a "tensor of contravariant rank 1 and covariant rank 1".

Exercise 3.7.4. Show that $\overline{W}^j_{;j} = W^j_{;j}$ (summation over j).

According to Exercise 3.7.4 we may define the (*covariant*) *divergence of* W, denoted div W, by

$$\text{div } W = W^j_{;j} \tag{46}$$

in any coordinate patch. Observe that whenever the Christoffel symbols are zero (e.g., in $\mathbb{R}^n, \mathcal{M}$, etc.) this reduces to the usual definition of divergence.

Exercise 3.7.5. Show that if $\{e_1,...,e_k\}$ is any orthonormal basis for $T_p(M)$, then at p

$$\text{div } W = \sum_{i=1}^{k} \varepsilon_i \langle D_{e_i} W, e_i \rangle ,$$

where $\varepsilon_i = \langle e_i, e_i \rangle$.

The special case of interest to us is described as follows: Let M be a (globally hyperbolic) spacetime with Cauchy time function T and let $S = T^{-1}(r)$ be one of the corresponding Cauchy surfaces. We know that the gradient of T is a smooth vector field on M which at each point of S is normal to S (Lemma 3.3.16). Moreover, ∇T is always timelike (and can be assumed past-directed) so that, in particular, $g(\nabla T, \nabla T)$ is never zero. Thus, we may define a future-directed unit timelike vector field N on M by

$$N = -|g(\nabla T, \nabla T)|^{-1/2} \nabla T .$$

Observe that, on S, N is a *unit normal field.* We intend to measure the expansion of the universe by the extent to which this unit normal field on S is "diverging". Specifically, we define the *mean curvature H_S of S in M* by

$$H_S = (\text{div } N)|_S .$$

Exercise 3.7.6. Let W be a smooth vector field on Einstein-deSitter spacetime $\mathcal{E}$. Show that, relative to the usual coordinate patch on $\mathcal{E}$,

$$\text{div } W = \frac{\partial W^1}{\partial u^1} + \frac{\partial W^2}{\partial u^2} + \frac{\partial W^3}{\partial u^3} + \frac{2}{u^4} W^4 .$$

Using $T(u^1, u^2, u^3, u^4) = u^4$ as the global time function and denoting by S one of its Cauchy surfaces, say, $u^4 = u_0^4$ (a positive constant) show that

$$H_S(p) = H_S(u^1, u^2, u^3, u_0^4) = 2/u_0^4 .$$

Thus, the mean curvature of each $u^4 = u_0^4$ spacelike slice of $\mathcal{E}$ is a positive constant which becomes infinite as $u_0^4 \to 0$, i.e., as one approaches the "big bang".

The hypothesis of Hawking's theorem which corresponds to the physical assertion that the universe is expanding is the assumption that our given (globally hyperbolic) spacetime M contains at least one Cauchy surface on which the mean curvature is bounded below by some positive real number.

3.8. Singular Spacetimes and the Statement of Hawking's Theorem

The theorem of Hawking [H] to which we refer says, in effect, that any mathematical model of the event world (spacetime) which contains no causal anomalies (stably causal) and is deterministic (globally hyperbolic) and which models an expanding universe (positive lower bound on the mean curvature of one of its spatial cross-sections) in which gravity is always attractive (Ric $(V,V) \geq 0$ for all timelike V) must be "singular". All we lack for a careful statement of the theorem is the precise sense in which a spacetime can be "singular". Ordinarily, of course, a "singularity" is a "point where something goes wrong". For example, in Chapter 2 an electromagnetic field (e.g., the Coulomb field) on $\mathcal{M}$ was taken to be singular at some point if one or more of its components relative to admissible coordinates became infinite there. In our present circumstances, however, we have no collection of distinguished coordinate systems to use as a benchmark and it is often very difficult to decide whether a given "singularity" is "real" or only due to an unfortunate choice of coordinates.

Exercise 3.8.1. Let $D = \{(u^1,u^2,u^3,u^4) \in \mathbb{R}^4 : u^4 > 0\}$ and suppose that for some spacetime M there is a coordinate patch $\chi: D \to M$ relative to which the nonzero metric components are $g_{11} = g_{22} = g_{33} = 1$ and $g_{44} = -\dfrac{1}{(u^4)^2}$. Observe that $g_{44} \to -\infty$ as $u^4 \to 0$. Show that, nevertheless, M is just a disguised version of $\mathcal{M}$.

There is an even more fundamental difficulty, however, since the "field" we have in mind here is the metric g itself and, by definition, the metric on a spacetime must be well-behaved (smooth) at each point of the underlying manifold. If the object before us is indeed a spacetime, then any potential "singular points" must already have been cut out. The questions then is "How does one detect the 'holes' that remain after the bad points were removed?" The answer is, at least on the surface, almost childishly simple: One knows there is a hole if something falls through it! Somewhat more precisely, we shall take as an indication that a region has been excised from our spacetime M the existence of a timelike geodesic (freely falling material particle) whose domain of definition cannot be extended to all of $\mathbb{R}$. The motivation is as follows: By Theorem 3.3.3 an inextendible geodesic cannot have endpoints since, at each of its points, it has a well-defined velocity vector which can be used to extend the geodesic to slightly larger (or smaller) values of its parameter. If one proceeds in this way to extend the geodesic to larger and larger (or smaller and smaller) values of its parameter one cannot fail to succeed unless the extensions become "shorter and shorter" and run out of the manifold. Thus, a timelike geodesic on M which is not defined on all of $\mathbb{R}$ is either extendible or "falls out" of the manifold in some finite amount of proper time. From the physical point of view this last possiblity is as "singular" as anything one might imagine since it indicates the presence of a freely falling material particle which simply ceases to exist after some finite lapse of its own proper time. The precise definitions follow.

A manifold M with metric is said to be *geodesically complete* if each of its inextendible geodesics is defined on all of $\mathbb{R}$.

Exercise 3.8.2. Show that M is geodesically complete if and only if for every $p \in M$ the exponential map $\exp_p$ is defined on all of $T_p(M)$.

A spacetime M is *timelike* (resp., *null, spacelike*) *geodesically complete* if each of its inextendible timelike (resp., null, spacelike) geodesics is defined on all of $\mathbb{R}$; otherwise, M is *timelike* (resp., *null, spacelike*) *geodesically incomplete.*

Remarks. For our purposes here a spacetime is "singular" if and only if it is timelike geodesically incomplete. Nevertheless, one must recognize that there are other ways, equally objectionable from the physical point of view, for a spacetime to deserve the appellation "singular". Null geodesic incompleteness is certainly one. On the other hand, the physical interpretation of spacelike geodesic incompleteness is not so clear. However, since accelerated observers seem to have as much right as free observers to object to their existence being abruptly curtailed one might attempt to generalize our notion of incompleteness to include a wider class of timelike curves. More information on these various possibilities is availabe in [HE]. One final remark. Questions concerning geodesic completeness in Lorentzian manifolds can be quite delicate and subtle, but for Riemannian manifolds the situation is well understood. Indeed, there is a classical theorem of Hopf and Rinow (see [O2]) which asserts that a Riemannian manifold is geodesically complete if and only if it is complete as a metric space (Cauchy sequences converge). Here the metric is the natural distance function derivable from the Riemannian metric (the distance between two points is the infimum of the arc lengths of all curves joining them in the manifold); see Problem 4.C.

$\mathcal{M}$ and $\mathcal{D}$ are, of course, geodesically complete, but $\mathcal{E}$ is not since the "vertical" timelike geodesics $\mu(t) = (u_0^1, u_0^2, u_0^3, t)$, $0 < t < \infty$, cannot be extended to values of $t \le 0$

(the scalar curvature S of $\mathcal{E}$ becomes infinite along such a geodesic as $t \to 0$ and so would have to take the value ∞ at any point on the geodesic corresponding to $t = 0$). Einstein-deSitter spacetime is "singular" in our sense. Intuitively, the situation is something like this: In $\mathcal{E}$ the universe is expanding, i.e., distances between galaxies increase as $u^4 \to \infty$. Consequently, as one proceeds backward in time, i.e., lets $u^4 \to 0$, the universe contracts toward the situation in which all of its mass is concentrated at a single "point" at which our ability to make measurements breaks down completely (the scalar and mean curvatures become infinite there). Note, however, that the "singularity" is not really a point in $\mathcal{E}$, but rather corresponds to the entire "missing" hyperplace $u^4 = 0$. It was the occurrence of such singular behavior in many of the exact solutions to the field equations which first suggested that this might be a characteristic feature of general relativity. The singularity theorems of Hawking and Penrose are generally regarded as confirmation of this suspicion.

The reader may have noticed an apparent weakness in our use of timelike geodesic incompleteness as the criterion for "singularity". It would seem that even the most innocuous of spacetimes (e.g., $\mathcal{M}$) can be made "singular" according to our definition by simply deleting a single point. In fact, this is true and one surely would not, on physical grounds want to regard, say, $\mathcal{M} - \{(0,0,0,0)\}$ as singular in any sense. How does one distinguish between "holes" which must be there because the spacetime is "really singular" and those which are capriciously plucked out of a perfectly respectable manifold? We propose that the difference between $M = \mathcal{M} - \{(0,0,0,0)\}$ and $\mathcal{E}$ is that M is a submanifold of a larger spacetime in which the hole is filled in, but $\mathcal{E}$ is not (again because the scalar curvature of $\mathcal{E}$ blows up as $u^4 \to 0$). More precisely, we say that a spacetime is *maximal* if it is not a proper submanifold of any other spacetime. $\mathcal{E}$ is maximal (see Exercise 5.2.7 of [SW2]), but $\mathcal{M} - \{(0,0,0,0)\}$ is not. Geodesic incompleteness is of geometrical and physical significance only for maximal spacetimes.

Finally then we are in a position to state the theorem to which all of our efforts have been directed. The proof will be given in Chapter 4.

Theorem 3.8.1. (Hawking [H]): Let M be a (time orientable, stably causal, globally hyperbolic)spacetime which satisfies

1. Ric $(V,V) \geq 0$ for all timelike tangent vectors V, and
2. There exists a Cauchy surface S in M on which the mean curvature H_S is bounded below by some positive constant k, i.e., $H_S(p) \geq k$ for each p in S.

Then M is timelike geodesically incomplete. More precisely, if $\mu: (-u_0, 0] \to M$ is any future-directed timelike geodesic such that $g(\mu', \mu') = -1$, $\mu(0) \in S$ and $\mu'(0)$ is normal to S in M, then $-u_0 \geq -3/k$.

PROBLEMS

3.A. Geodesics of the Sphere

Give an argument analogous to that we employed in section 3.3 for calculating the geodesics of deSitter spacetime to show that the geodesics of the 2-spheres S^2 are the constant speed parametrizations of its great circles (intersections with S^2 of planes in $\mathbb{R}^3$ through the origin). Generalize to S^n.

3.B. Curves in $\mathbb{R}^3$

We denote by $<,>$ the usual Riemannian metric (dot product) on $\mathbb{R}^3$ and by x,y and z the standard coordinates. For any smooth curve $\alpha: I \to \mathbb{R}^3$, $\alpha(t) = (x(t), y(t), z(t))$, one defines an *arc length parameter* $s = s(t)$ by

$$s = s(t) = \int_a^t <\alpha'(t), \alpha'(t)>^{1/2} dt$$

$$= \int_a^t ((x'(t))^2 + (y'(t))^2 + (z'(t))^2)^{1/2} dt$$

(this arc length parameter is said to be *based at a* in I). We will assume that all of our curves α have been reparametrized in terms of arc length so that $\alpha = \alpha(s)$.

1. Describe the smooth curve $\alpha(t) = (\cosh t, \sinh t, t)$, $t \in \mathbb{R}$, in $\mathbb{R}^3$ and show that its arc length reparametrization (based at 0) is $\alpha(s) = ((1 + s^2/2)^{1/2}, s/\sqrt{2}, \sinh^{-1}(s/\sqrt{2}))$.

2. Show that any curve $\alpha = \alpha(s)$ parametrized by arc length has *unit speed*, i.e., $<\alpha'(s), \alpha'(s)> = 1$ for all s.

Henceforth we denote $\alpha'(s)$ by $T(s)$ and refer to it as the *unit tangent* field to α. $T'(s) = \alpha''(s)$ is the *curvature vector field* along α and its magnitude $k(s) = \| T'(s) \| = <T'(s), T'(s)>^{1/2}$ is called the *curvature* of α.

3. Show that T' is orthogonal to T, i.e. $<T,T'> = 0$. *Hint.* Differentiate $T(s) \cdot T(s) = 1$.

4. Observe that $k(s) \geq 0$ and show that if $k(s) = 0$ for some interval of s values, then α is a linearly parametrized straight line on that interval.

Henceforth we assume that $k(s) \neq 0$ on the interval of s values under consideration. Define that *principal normal* vector field along α by $N(s) = \frac{1}{k(s)} T'(s)$ and the *binormal* field by $B(s) = T(s) \times N(s)$. The 2-dimensional affine plane at $\alpha(s)$ spanned by $T(s)$ and $N(s)$ is the *osculating plane* at $\alpha(s)$.

Figure 3.13

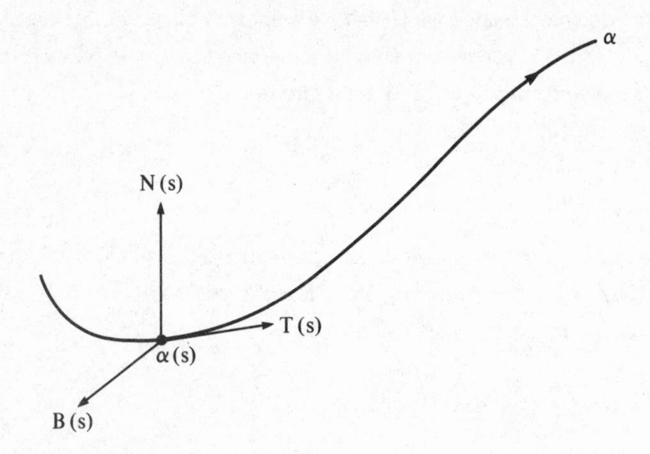

The triple $\{T(s), N(s), B(s)\}$ of vector fields along α is called the *Frenet frame* of α and gives an orthonormal basis for $T_{\alpha(s)}(\mathbb{R}^3)$ at each $\alpha(s)$.

5. Prove the *Frenet formulas*:

$$T'(s) = k(s)\,N(s)$$

$$N'(s) = -k(s)\,T(s) + \tau(s)\,B(s)$$

$$B'(s) = -\tau(s)\,N(s)$$

where $\tau(s) = -<B(s), N(s)>$ is the *torsion* of α.

6. Calculate the Frenet frame for the unit speed helix $\alpha(s) = (a\cos(s/c),\ a\sin(s/c),\ bs/c)$, where $c = (a^2 + b^2)^{1/2}$ and $a > 0$ and show that $k(s) = \frac{a}{a^2 + b^2}$ and $\tau(s) = \frac{b}{a^2 + b^2}$.

7. Let $\alpha(s)$ be a unit speed curve in $\mathbb{R}^3$ with $k(s) > 0$. Show that α is a plane curve (i.e., lies entirely in some plane in $\mathbb{R}^3$) if and only if $\tau(s) = 0$ for all s. *Hint.* Show that if α is a plane curve, then $B(s)$ is constant. Conversely, assume $\tau(s) = 0$ and compute $\frac{d}{ds}(\alpha(s) - \alpha(0)) \bullet B(s)$ to show $(\alpha(s) - \alpha(0)) \bullet B = 0$.

8. Show that if $\alpha(s)$ is a unit speed curve with constant positive curvature k and $\tau(s) = 0$ for all s, then α is part of a circle of radius $1/k$. *Hint.* Show that the curve $\gamma(s) = \alpha(s) + \frac{1}{k} N(s)$ is actually constant, i.e., $\gamma(s) = p$ for every s. p is the center of the circle.

Remark. Many more such applications of the Frenet formulas to the theory of curves in $\mathbb{R}^3$ are available in [O1].

3.C. Algebraic Symmetries of the Curvature Tensor

We define the covariant components of the Riemann curvature tensor R^a_{bcd} by

$$R_{abcd} = g_{a\alpha} R^\alpha_{bcd}, \quad a,b,c,d = 1,\ldots,k \ .$$

1. Prove that

$$R_{abcd} = \frac{1}{2}\left[\frac{\partial^2 g_{ac}}{\partial u^d \partial u^b} - \frac{\partial^2 g_{bc}}{\partial u^d \partial u^a} - \frac{\partial^2 g_{ad}}{\partial u^c \partial u^b} + \frac{\partial^2 g_{bd}}{\partial u^c \partial u^a}\right]$$

$$+ g_{\alpha\beta}\left[\Gamma^\alpha_{ca}\Gamma^\beta_{bd} - \Gamma^\alpha_{da}\Gamma^\beta_{bc}\right]$$

2. Derive the following symmetries of the curvature tensor:

$$R_{abcd} = R_{cdab}$$

$$R_{abcd} = -R_{bacd} = -R_{adbc} = R_{badc}$$

$$R_{abcd} + R_{adbc} + R_{acbd} = 0$$

3.D. Some Surface Theory

Throughout this problem $\chi : D \to M$ will be a coordinate patch for the smooth surface (2-manifold) M in $\mathbb{R}^3$ with D a connected open subset of the plane $\mathbb{R}^2$. The Riemannian metric on M is the restriction of the $\mathbb{R}^3$-inner product to each tangent plane. All curves in M will be parametrized by arc length (see Problem 3.B) and the unit normal vector field on $\chi(D)$ is taken to be $U = \chi_1 \times \chi_2 \,/\, \| \chi_1 \times \chi_2 \|$.

1. Let $g = \det(g_{ij})$ be the determinant of the metric tensor for M. Show that $\|\chi_1 \times \chi_2\|^2 = g$.

As in Exercise 3.3.2 we resolve the acceleration (curvature) α'' of an arbitrary smooth unit speed curve α in M into tangential and normal components: $\alpha'' = \alpha''_{\text{tan}} + \alpha''_{\text{nor}}$, where $\alpha''_{\text{nor}} = L_{ij}\,(u^i)'\,(u^j)'U$, $L_{ij} = \chi_{ij} \cdot U$, and $\alpha''_{\text{tan}} = ((u^r)'' + \Gamma^r_{ij}\,(u^i)'\,(u^j)')\,\chi_r$. The matrix (L_{ij}) is called the *second fundamental form* of M ((g_{ij}) is often called the *first fundamental form* of M). For any *unit* tangent vector $\mathrm{v} = \mathrm{v}^i\,\chi_i$ in $T_p(M)$ the *normal curvature of M at p in the direction* v is

$$k_N(\mathrm{v}) = L_{ij}(p)\mathrm{v}^i\mathrm{v}^j \ .$$

2. Let M be the graph of $z = f(x,y) = \frac{1}{2}(y^2 - x^2)$ (a saddle). More precisely, define $\chi : \mathbb{R}^2 \to \mathbb{R}^3$ by $\chi(u,\mathrm{v}) = (u, \mathrm{v}, \frac{1}{2}(\mathrm{v}^2 - u^2))$ and set $M = \chi(\mathbb{R}^2)$. Calculate $(L_{ij}(u,\mathrm{v}))_{i,j=1,2}$ and show that, at the origin, $(L_{ij}(0,0))$ is

$$\begin{bmatrix} -1 & 0 \\ 0 & 1 \end{bmatrix} .$$

Observe that $T_{(0,0)}(M)$ is just the xy-plane so that any unit vector v in $T_{(0,0)}(M)$ can be written $\mathrm{v} = (\cos\theta)\,\chi_1 + (\sin\theta)\,\chi_2$, where θ is the usual polar angle. Show that

$$k_N(\mathrm{v}) = -\cos 2\theta \ .$$

3. Show that for an arbitrary M and unit vector v in $T_p(M)$,

$$k_N(\mathrm{v}) = \alpha'' \cdot U \ ,$$

where α is any unit speed curve in M that fits v at p.

4. We gain some intuitive appreciation of normal curvature by applying the result of #3 to certain specific curves in M. Fix a p in M and a unit vector v in $T_p(M)$. The plane through p spanned by v and $U(p)$ intersects M in a curve whose unit speed parametrization is denoted $\alpha_{\mathrm{v}} = \alpha_{\mathrm{v}}(s)$ and called the *normal section of M at p in the direction* v. Show that $k_N(\mathrm{v}) = \pm\,\|\alpha''_{\mathrm{v}}\|$ (+ if α''_{v} is in the direction of U and – if α''_{v} is in the direction of $-U$). Thus, k_N keeps track of both the magnitude and direction of M's curvature in every direction (see Problem 3.B).

5. Show that there exist unit vectors v_1 and v_2 in $T_p(M)$ (called the *principal directions* of M at p) such that $k_1 = k_N(v_1)$ and $k_2 = k_N(v_2)$ are respectively the maximum and minimum values of $k_N(v)$ at p (k_1 and k_2 are called the *principal curvatures* of M at p). *Hint.* Regard k_N as a real-valued function on the unit circle in $T_p(M)$.

The *Gaussian curvature* $K = K(p)$ of M at p is defined by $K(p) = k_1 k_2$, where k_1 and k_2 are the principal curvatures of M at p.

6. Let M be the saddle in #2 and $p = (0,0)$. Show that the Gaussian curvature of M at p is -1.

7. Use the algebraic symmetries of the curvature tensor from Problem 3.C to show that, for a surface, the only nonzero components of R_{abcd} are R_{1212}, R_{2121}, R_{1221}, and R_{2112} and that these satisfy $R_{1212} = R_{2121} = -R_{2112} = -R_{1221}$.

Thus, in dimension two, the curvature tensor has essentially only one independent component. Gauss' famous *Theorema Egregium* states that, in fact, $R_{1212} = K_g$ and implies, in particular, that the Gaussian curvature can be defined entirely in terms of the metric (g_{ij}) of the surface. Remarkably, the function K which seems to be telling us how M "curves in $\mathbb{R}^3$" can be computed from measurements made entirely in M without any reference to the ambient Euclidean space $\mathbb{R}^3$. There is an elementary and very readable proof of this in [Fa].

3.E. Derivatives of Smooth Maps

Let M and N be smooth manifolds of dimension m and n respectively and $f : M \to N$ a smooth map. For each p in M we define a map

$$f_{*p} : T_p(M) \to T_{f(p)}(N)$$

as follows: For each v in $T_p(M)$ select a smooth curve α which fits v at p. Then $f \circ \alpha$ is a smooth curve in N which goes through $f(p)$ at $t = 0$. We set

$$f_{*p}(v) = (f \circ \alpha)'(0) .$$

f_{*p} is called the *derivative* (or *differential* or *tangent map*) *of f at p*.

1. Let $\chi : D \to M$ be a coordinate patch for M at p and let $\mathcal{Y} = f \circ \chi$ (which need *not* be a coordinate patch for N at $f(p)$). Define $\mathcal{Y}_i$ as usual by $\mathcal{Y}_i = \partial \mathcal{Y} / \partial u^i$, where $u^1, \ldots, u^m$ are the coordinates in $D \subseteq \mathbb{R}^m$. Show that

$$(f \circ \alpha)'(t) = (u^i(t))' \, \mathcal{Y}_i(u^1(t), \ldots, u^m(t)) .$$

2. Show that the definition of f_{*p} does not depend on the choice of α.
3. Show that $f_{*p} : T_p(M) \to T_{f(p)}(N)$ is a linear transformation.
4. Choose a coordinate patch χ for M near p and a coordinate patch Z for N near $f(p)$ and express f in terms of these coordinates, i.e., consider the map $Z^{-1} \circ f \circ \chi$:

$$(z^1, ..., z^n) = (f^1(u^1, ..., u^m), ..., f^n(u^1, ..., u^m)) .$$

Show that the matrix of f_{*p} relative to the coordinate bases for χ and Z is the Jacobian

$$\begin{bmatrix} \frac{\partial f^1}{\partial u^1} & \cdots & \frac{\partial f^1}{\partial u^m} \\ \cdot & & \cdot \\ \cdot & & \cdot \\ \cdot & & \cdot \\ \frac{\partial f^n}{\partial u^1} & \cdots & \frac{\partial f^n}{\partial u^m} \end{bmatrix}$$

5. Show that if $f: M \to N$ and $g: N \to P$ are smooth and p is in M, then $(g \circ f)_{*p} = g_{*f(p)} \circ f_{*p}$ and explain why this fact is called the *Chain Rule*.
6. Suppose that $f: M \to N$ has the property that f_{*p} is a linear isomorphism. Show that f is a local diffeomorphism at p, i.e., that f maps some open neighborhood of p in M diffeomorphically onto an open neighborhood of $f(p)$ in N.
7. Prove that the converse of #6 is also true.
8. Prove that if M is diffeomorphic to N, then $m = n$.

Remark. The corresponding topological result (i.e., that $\mathbb{R}^m$ homeomorphic to $\mathbb{R}^n$ implies $m = n$) is *much* deeper and requires considerable machinery to prove; see [Nab].

3.F. Fermi-Walker Transport

Parallel translation along a smooth curve α preserves dot products (Theorem 3.3.4). Consequently, if one is given an orthonormal basis for the tangent space at one point $\alpha(t_0)$ it can be parallel translated to give orthonormal bases at each point along α. Such an ensemble of vector fields along α which give orthonormal bases for each $T_{\alpha(t)}(M)$ is called a *moving frame* along α and is somewhat analogous to the Frenet frames considered in problem 3.B. The major defect of such frames is that even if the velocity vector $\alpha'(t_0)$ is a member of the initial basis at $\alpha(t_0)$ the velocity vector $\alpha'(t)$ will generally not be in any of the translated bases (unless α is a geodesic). In this problem we consider another method of transporting vectors along smooth timelike curves in a spacetime which always preserves velocity vectors. Thus, we consider a smooth timelike curve $\alpha = \alpha(\tau)$ parametrized by proper time in a spacetime M. $\alpha'(\tau)$ is its unit tangent vector and $A = D_{\alpha'}\alpha'$ its covariant acceleration. If V is any smooth vector field along α we define its *Fermi-Walker derivative* along α by

$$F_{\alpha'}V = D_{\alpha'}V + \langle V, \alpha' \rangle A - \langle A, V \rangle \alpha' .$$

We say that V is *Fermi-Walker transported* along α if $F_{\alpha'} V = 0$.

1. Show that if α is geodesic, then $F_{\alpha'}V = D_{\alpha'}V$ for any V.
2. Show that α' is Fermi-Walker transported along α.
3. Let V and W be smooth vector fields along α. Show that $\dfrac{d}{d\tau} \langle V,W \rangle = \langle F_{\alpha'}V, W \rangle + \langle V, F_{\alpha'}W \rangle$ and conclude that Fermi-Walker transport preserves dot products.
4. Show that if V is orthogonal to α', then $F_{\alpha'}V$ is the projection of $D_{\alpha'}V$ onto the orthogonal complement of α'.

3.G Conformally Equivalent Spacetimes

Two spacetimes (M,g) and (M,g') with the same underlying manifold M are said to be *conformally equivalent* if $g = \Omega g'$, where $\Omega : M \to (0,\infty)$ is some positive smooth function on M.

1. Show that Einstein-deSitter spacetime $\mathcal{E}$ is conformally equivalent to the open submanifold $\mathbb{R}^3 \times (0,\infty)$ of Minkowski spacetime $\mathcal{M}$. *Hint.* Introduce a new time coordinate $u = 3(u^4)^{1/3}$ on $\mathcal{E}$ and show that the metric components relative to (u^1, u^2, u^3, u) are

$$g_{ij}(u^1,u^2,u^3,u) = \begin{cases} 0 & \text{if } i \neq j \\ (u/3)^4 & \text{if } i = j = 1,2,3 \\ -(u/3)^4 & \text{if } i = j = 4 \end{cases}$$

2. Suppose that (M,g) and (M,g') are conformally equivalent. Show that they have the same null curves and the same causal structure (i.e., $p \ll q$ is and only if $p \ll' q$). Must they have the same geodesics?
3. Show that any spacetime conformally equivalent to a time orientable spacetime is itself time orientable. Is stable causality also a conformal invariant? Global hyperbolicity?

CHAPTER 4

THE PROOF OF HAWKING'S THEOREM

4.1. A Sketch of the Argument

In a globally hyperbolic spacetime M with Cauchy surface S any point p in $M-S$ is connectible to S by (in general, many) smooth timelike curves. Each such curve has a proper time "distance" from S. Hawking's Theorem depends in a crucial way upon the fact that among these curves there is a least one geodesic that "maximizes" the proper time distance from p to S.

Theorem 4.1.1. Let M be a (globally hyperbolic) spacetime, S a Cauchy surface in M and p a point in $M-S$. Then there exists at least one smooth timelike curve $\lambda:[0,1] \to M$ such that $\lambda(0) = p$, $\lambda(1) \in S$ and the proper time length $\int_0^1 | g(\lambda'(t), \lambda'(t)) |^{1/2}\, dt$ of λ is at least as large as the proper time length of any other smooth timelike curve from p to (a point in) S. Moreover, this curve is a geodesic which hits S orthogonally ($g(\lambda'(1), \mathrm{v}) = 0$ for every v in $T_{\lambda(1)}(S)$).

Theorem 4.1.1 is quite deep and we shall defer any discussion of its proof until somewhat later. With it and a bit more technical machinery, however, the proof of Hawking's Theorem is within reach. Roughly, the argument goes something like this (we shall employ the notation established in Theorem 3.8.1): $\mu: (-u_0, 0] \to M$ is a future-directed timelike geodesic such that $g(\mu', \mu') = -1$, $\mu(0) \in S$ and $\mu'(0)$ is normal to S. Fix an arbitrary $-\mathrm{v}_0$ in $(-u_0, 0]$. We show that $-\mathrm{v}_0 \geq -3/k$ and conclude that $-u_0 \geq -3/k$. We shall denote by p the point $\mu(-\mathrm{v}_0)$.

Exercise 4.1.1. Use Theorem 4.1.1 to show that there exists a future-directed timlike geodesic $\gamma: [-a, 0] \to M$ (a positive real number) such that $\gamma(-a) = p$, $\gamma(0) \in S$, $\gamma'(0)$ is orthogonal to S, $g(\gamma', \gamma') = -1$ and such that the proper time length a of γ is greater than or equal to the proper time length of any other smooth timelike curve in M from p to a point in S.

Since a is the proper time length of γ from p to S and v_0 is that of μ, maximality implies that $v_0 \leq a$ and it will suffice to show that $-a \geq -3/k$.

Figure 4.1

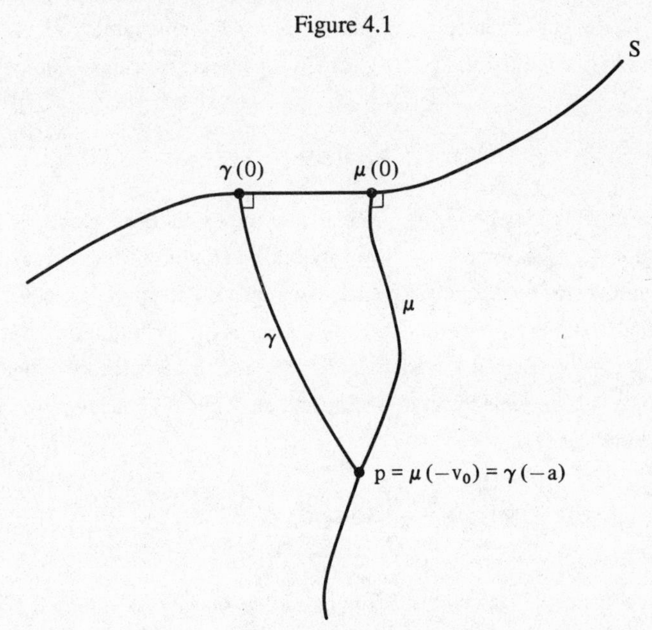

Now we forget about μ and concentrate on γ. Maximality implies that any timelike "variation" of γ which connects p and S must be "shorter". Somewhat more precisely, we consider a 1-parameter family of smooth timelike curves $\tau^v: [-a, 0] \rightarrow M$, $-\delta < v < \delta$, such that, for each v, $\tau^v(-a) = p$, $\tau^v(0) \in S$ and $\tau^0 = \gamma$.

Figure 4.2

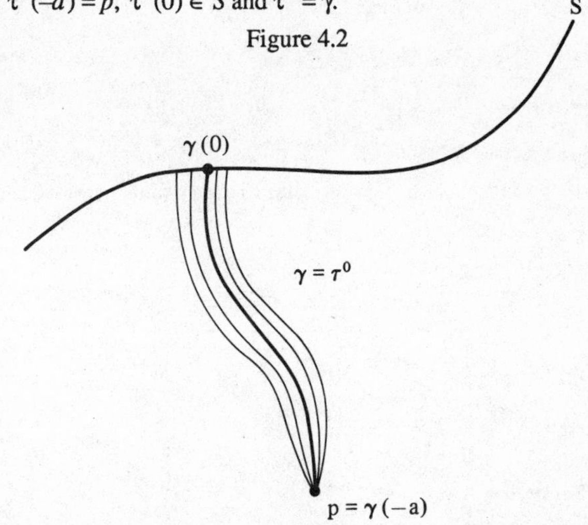

Define a real-valued function L on $(-\delta,\delta)$ by setting $L(\mathrm{v})$ equal to the proper time length of τ^{v} between p and S, i.e.,

$$L(\mathrm{v}) = \int_{-a}^{0} |g((\tau^{\mathrm{v}})', (\tau^{\mathrm{v}})')|^{1/2}\, dt \ .$$

Thus, $L(0)$ is the length of γ and, by assumption, is a maximum value for $L: (-\delta,\delta) \to \mathbb{R}$. Consequently, the second derivative $L''(\mathrm{v})$ of L with respect to v must non-positive at 0, i.e.,

$$L''(0) \leq 0 \ . \tag{1}$$

As it happens there is a well-known formula from the calculus of variations (which we derive in section 4.2) that expresses $L''(0)$ for such a variation of γ in terms of γ, the so-called "variation vector field" along γ (which essentially points in the direction in which γ is being "varied") and the Riemann curvature tensor of M. Writing down the inequality (1) for three carefully selected variations of γ (one in each of three orthogonal spacelike directions along γ), using this formula, adding and algebraically manuvering a bit eventually yields the inequality

$$0 \leq \frac{3}{a} - \int_{-a}^{0} \left(\frac{a+u}{a}\right)^{2} \mathrm{Ric}\,(\gamma', \gamma')\, du - H_S\,(\gamma(0)) \ . \tag{2}$$

One of the hypotheses of Hawking's Theorem 3.8.1 is that $\mathrm{Ric}\,(V,V) \geq 0$ for all timelike V. Since γ is timelike the integral in (2) is non-negative so

$$0 \leq \frac{3}{a} - H_S(\gamma(0)) \ . \tag{3}$$

Another hypothesis of our theorem is that the mean curvature H_S of S in M is bounded below by some $k > 0$ so that $H_S(\gamma(0)) \geq k$ and therefore (3) gives us $0 \leq \frac{3}{a} - k$, i.e., $-a \geq -3/k$ as required.

In the next few sections we develop more carefully the notion of a "variation" of the curve γ, derive the variational formula referred to above and then give a detailed proof of Theorem 3.8.1.

4.2. Two-Parameter Maps and the Variational Formulas

We consider a smooth k-manifold M with metric. A *two-parameter map* in M is a smooth mapping τ of some rectangle $[a,b] \times (-\delta,\delta)$ into M. The *u- and v-parameter curves* of τ are respectively the curves $u \to \tau(u,\mathrm{v})$ (v fixed) and $\mathrm{v} \to \tau(u,\mathrm{v})$ (u fixed). A *vector field on* τ is a mapping Z which assigns to each (u,v) in $[a,b] \times (-\delta,\delta)$ a tangent vector $Z(u,\mathrm{v})$ in $T_{\tau(u,\mathrm{v})}(M)$. Z is *smooth* if for every coordinate patch χ on M the component functions $Z^i(u,\mathrm{v})$ defined by $Z(u,\mathrm{v}) = Z^i(u,\mathrm{v})\,\chi_i(\tau(u,\mathrm{v}))$ are C^∞. The most obvious examples of smooth vector fields on τ are the *coordinate velocity vector fields* τ_u and τ_v which assign to every (u,v) the velocity vector of the u-(respectively, v-) parameter curve through $\tau(u,\mathrm{v})$. For each χ we write $\chi^{-1} \circ \tau(u,\mathrm{v}) = (u^1(u,\mathrm{v}),\ldots,u^k(u,\mathrm{v}))$ so that $\tau_u = \dfrac{\partial u^i}{\partial u}\,\chi_i$ and similarly for τ_v. Restricting any smooth vector field Z on τ to a parameter curve (i.e., holding either u or v fixed) gives a smooth vector field along that curve.

If $\gamma\colon [a,b] \to M$ is a smooth curve in M, then the two-parameter map $\tau : [a,b] \times (-\delta,\delta) \to M$ is a *variation* of γ if γ is the u-parameter curve corresponding to $\mathrm{v} = 0$, i.e., if $\gamma(u) = \tau(u,0)$ for each u in $[a,b]$.

Figure 4.3

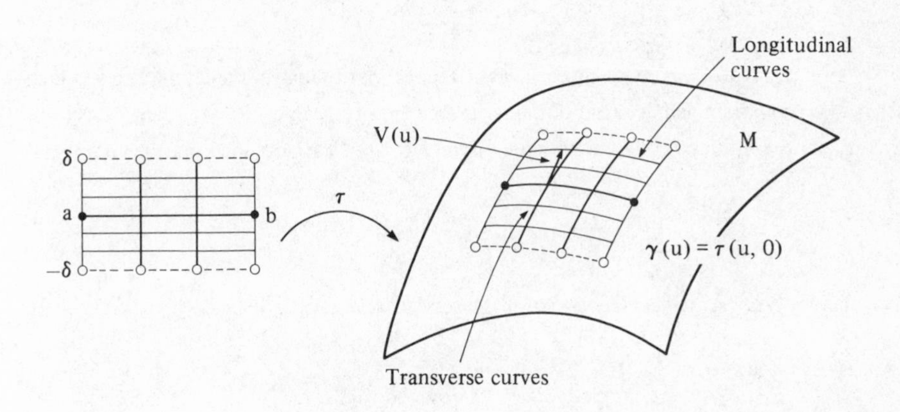

In this case the u-parameter curves are called the *longitudinal curves* of τ and are regarded as small variations (deformations) of γ. The v-parameter curves are the *transverse curves* of τ and are thought of as connecting "nearby" variations of γ. The vector field along γ

obtained by attaching to each point in the image of γ the velocity vector of the transverse v-parameter curve through that point, i.e., by setting

$$V(u) = \tau_v(u, 0)$$

is called the *variation vector field* of τ and is thought of as "pointing toward nearby" variations of γ.

Exercise 4.2.1. Show that every smooth vector field V along γ is the variation vector field for some variation of γ. *Hint.* Consider $\exp_{\gamma(u)}(vV(u))$.

Now we return to an arbitrary smooth vector field Z on the two-parameter map τ. We define the *partial covariant derivatives* Z_u and Z_v to be the covariant derivatives of Z along the u- and v-parameter curves. More precisely, at each (u_0, v_0),

$$Z_u(u_0, v_0) = \text{covariant derivative of } u \to Z(u, v_0) \text{ at } u = u_0$$

and similarly for Z_v. Thus, if $Z = Z^i \chi_i$, then

$$Z_u = \left[\frac{\partial Z^r}{\partial u} + \Gamma^r_{ij} \frac{\partial u^j}{\partial u} Z^i \right] \chi_r$$

and similarly for Z_v. Two special cases are of particular interest. If $Z = \tau_u$, then $Z_u = \tau_{uu}$ = covariant acceleration of the u-parameter curves and if $Z = \tau_v$, then $Z_v = \tau_{vv}$ = covariant acceleration of the v-parameter curves. If τ is a variation of some curve γ, then

$$A(u) = \tau_{vv}(u, 0)$$

is called the *transverse acceleration* of τ. Observe that if $Z = \tau_u$, then $Z_v = \tau_{uv}$ and

$$\tau_{uv} = \left[\frac{\partial^2 u^r}{\partial v \partial u} + \Gamma^r_{ij} \frac{\partial u^j}{\partial v} \frac{\partial u^i}{\partial u} \right] \chi_r = \tau_{vu}$$

since Γ^r_{ij} is symmetric in i and j and mixed second order partial derivatives are equal. Thus, we have proved

Lemma 4.2.1. If τ is a two-parameter map in M, then $\tau_{uv} = \tau_{vu}$.

For an arbitrary smooth vector field Z on τ, however, $Z_{uv} - Z_{vu}$ will, in general, not be zero. Indeed, we now calculate $Z_{uv} - Z_{vu}$ relative to an arbitrary coordinate patch χ as follows: $Z_u = (Z_u)^r \chi_r$, where

$$(Z_u)^r = \frac{\partial Z^r}{\partial u} + \Gamma^r_{ij} \frac{\partial u^j}{\partial u} Z^i \ .$$

Exercise 4.2.2. Calculate $Z_{u\mathrm{v}} = (Z_u)_{\mathrm{v}}$ and show that its rth component is

$$(Z_{u\mathrm{v}})^r = \left[\frac{\partial^2 Z^r}{\partial \mathrm{v} \partial u} + \Gamma^r_{mn} \frac{\partial^2 u^n}{\partial \mathrm{v} \partial u} Z^m \right] + \Gamma^r_{mn} \frac{\partial Z^m}{\partial \mathrm{v}} \frac{\partial u^n}{\partial u} + \tag{4}$$

$$+ \frac{\partial u^n}{\partial u} Z^m \frac{\partial \Gamma^r_{mn}}{\partial \mathrm{v}} + \Gamma^r_{ij} \frac{\partial u^j}{\partial \mathrm{v}} \frac{\partial Z^i}{\partial u} + \Gamma^r_{ij} \Gamma^i_{mn} \frac{\partial u^j}{\partial \mathrm{v}} \frac{\partial u^n}{\partial u} Z^m$$

Observe that the term in brackets in (4) is symmetric in u and v so that the rth component of $Z_{u\mathrm{v}} - Z_{\mathrm{v}u}$ is

$$(Z_{u\mathrm{v}} - Z_{\mathrm{v}u})^r = \left[\frac{\partial \Gamma^r_{mn}}{\partial \mathrm{v}} \frac{\partial u^n}{\partial u} - \frac{\partial \Gamma^r_{mn}}{\partial u} \frac{\partial u^n}{\partial \mathrm{v}} + \Gamma^r_{ij} \Gamma^i_{mn} \frac{\partial u^j}{\partial \mathrm{v}} \frac{\partial u^n}{\partial u} \right. \tag{5}$$

$$\left. - \Gamma^r_{ij} \Gamma^i_{mn} \frac{\partial u^j}{\partial u} \frac{\partial u^n}{\partial \mathrm{v}} \right] Z^m$$

In order to save some space we introduce the notation $\frac{\partial u^1}{\partial \mathrm{v}} = \tau^1_{\mathrm{v}}$, $\frac{\partial u^n}{\partial u} = \tau^n_u$, etc. With this and the Chain Rule applied to the derivatives of the Christoffel symbols (5) becomes

$$(Z_{u\mathrm{v}} - Z_{\mathrm{v}u})^r = \left[\frac{\partial \Gamma^r_{mn}}{\partial u^1} \tau^1_{\mathrm{v}} \tau^n_u - \frac{\partial \Gamma^r_{mn}}{\partial u^1} \tau^1_u \tau^n_{\mathrm{v}} + \Gamma^r_{ij} \Gamma^i_{mn} \tau^j_{\mathrm{v}} \tau^n_u \right.$$

$$\left. - \Gamma^r_{ij} \Gamma^i_{mn} \tau^j_u \tau^n_{\mathrm{v}} \right] Z^m$$

$$= \left[\frac{\partial \Gamma^r_{mj}}{\partial u^n} \tau^n_{\mathrm{v}} \tau^j_u - \frac{\partial \Gamma^r_{mn}}{\partial u^j} \tau^n_{\mathrm{v}} \tau^j_u + \Gamma^r_{in} \Gamma^i_{mj} \tau^n_{\mathrm{v}} \tau^j_u \right.$$

$$\left. - \Gamma^r_{ij} \Gamma^i_{mn} \tau^n_{\mathrm{v}} \tau^j_u \right] Z^m$$

$$= \left[\frac{\partial \Gamma^r_{mj}}{\partial u^n} - \frac{\partial \Gamma^r_{mn}}{\partial u^j} + \Gamma^r_{in} \Gamma^i_{mj} - \Gamma^r_{ij} \Gamma^i_{mn} \right] \tau^n_{\mathrm{v}} \tau^j_u Z^m$$

$$= R^r_{mnj} \tau^n_{\mathrm{v}} \tau^j_u Z^m \ .$$

Before stating our result as a Theorem we introduce some standard notation and terminology. If X, Y and Z are tangent vectors at p in M, then $R(X,Y)Z = R_{XY}Z$ is the tangent vector at p defined relative to any coordinate patch χ by

$$R(X,Y)Z = R_{XY}Z = (R^a_{bcd}\, Z^b X^c Y^d)\, \chi_a \ . \tag{6}$$

If X, Y and Z are vector fields on M, then $R(X,Y)Z = R_{XY}Z$ is a vector field on M defined at each point p by (6). For fixed X and Y the operator $R(X,Y) = R_{XY}$ which takes Z to $R(X,Y)Z = R_{XY}Z$ is called the *curvature operator* of M. The result of our computation above now takes the following form.

Theorem 4.2.2. If Z is a smooth vector field on the two-parameter map τ on M, then

$$Z_{u\mathrm{v}} - Z_{\mathrm{v}u} = R(\tau_{\mathrm{v}}, \tau_u)Z = R_{\tau_{\mathrm{v}}\tau_u} Z \ . \tag{7}$$

Exercise 4.2.3. Show that if $\{e_1, \dots, e_k\}$ is any orthonormal basis for $T_p(M)$, then, at p,

$$\mathrm{Ric}\,(X,Y) = \sum_{i=1}^{k} \varepsilon_i g\,(R_{e_i X}Y,\ e_i) \ ,$$

where $\varepsilon_i = g\,(e_i, e_i)$.

Now we restrict our attention to the special case in which M is a spacetime, $\gamma : [a,b] \to M$ is a smooth future-directed timelike curve (parametrized by proper time so that $g\,(\gamma', \gamma') = -1$) and $\tau : [a,b] \times (-\delta, \delta) \to M$ is a variation of γ.

Exercise 4.2.4. Show that by taking δ sufficiently small we can insure that all of the longitudinal curves $u \to \tau(u, \mathrm{v}_0)$ are also timelike and future-directed. *Hint.* Restrict τ to a compact set of the form $[a,b] \times [-\delta_1, \delta_1]$, where $\delta_1 < \delta$ and argue by contradiction.

Now define a length function $L : (-\delta, \delta) \to M$ associated with τ by taking $L(\mathrm{v})$ to be the proper time length of the longitudinal curve $u \to \tau(u, \mathrm{v})$, $a \le u \le b$ (notice that u will, in general, be a proper time parameter only for the $\mathrm{v} = 0$ longitudinal curve, i.e., γ), i.e.,

$$L(\mathrm{v}) = \int_a^b |\, g\,(\tau_u(u,\mathrm{v}),\ \tau_u(u,\mathrm{v}))\,|^{1/2}\, du \tag{8}$$

for each v in $-\delta < \mathrm{v} < \delta$. Thus, $L(0)$ = length of $\gamma = b - a$. Since the longitudinal curves are all timelike, $g\,(\tau_u, \tau_u)$ is never zero so that the integrand in (8) is differentiable at $\mathrm{v} = 0$.

Theorem 4.2.3. If $\gamma : [a,b] \to M$ is a smooth, future-directed timelike curve in M parametrized by proper time and $\tau : [a,b] \times (-\delta,\delta) \to M$ is a future-directed timelike variation of γ (see Exercise 4.2.4), then

$$L'(0) = -\int_a^b g(\gamma'(u), V'(u))\, du \ ,$$

where $V(u) = \tau_v(u,0)$ is the variation vector field and $V'(u) = \tau_{vu}(u,0)$ is the covariant derivative of V along γ.

Proof. For this proof (and the next) we set $H(u,v) = (-g(\tau_u,\tau_u))^{1/2}$. From (8) we find that

$$L'(0) = \int_a^b \frac{\partial}{\partial v} H(u,v)\Big|_{v=0} du \ . \tag{9}$$

But

$$\frac{\partial}{\partial v} H(u,v) = \frac{1}{2}[-g(\tau_u,\tau_u)]^{-1/2} \frac{\partial}{\partial v}[-g(\tau_u,\tau_u)]$$

$$= -\frac{1}{2}[-g(\tau_u,\tau_u)]^{-1/2}\, 2g(\tau_u,\tau_{uv}) \text{ by Lemma 3.3.3}$$

$$= -[-g(\tau_u,\tau_u)]^{-1/2}\, g(\tau_u,\tau_{vu}) \text{ by Lemma 4.2.1.}$$

Setting $v = 0$ we get $\tau_u(u,0) = \gamma'(u)$ and $[-g(\tau_u,\tau_u)]^{-1/2} = [-g(\gamma'(u),\gamma'(u))]^{-1/2} = 1$ so that

$$\frac{\partial}{\partial v} H(u,v)\Big|_{v=0} = -g(\gamma'(u), V'(u)) \ .$$

Integrating from a to b gives the result. Q.E.D.

As we pointed out in section 4.1 we are primarily interested in the second derivative $L''(0)$ and that only when γ is a geodesic and V is orthogonal to γ'. Next we shall derive a formula general enough for our purposes.

Theorem 4.2.4. Let $\gamma : [a,b] \to M$ be a future-directed timelike geodesic in M. Suppose $\tau : [a,b] \times (-\delta,\delta) \to M$ is a future-directed timelike variation of γ whose variation vector field V is orthogonal to γ, i.e., $g(\gamma',V) = 0$. If $L : (-\delta,\delta) \to \mathbb{R}$ is the length function of τ, then

$$L''(0) = \int_a^b (g(R_{V\gamma'}\gamma', V) - g(V', V')\}\, du - g(\gamma', A)\Big|_a^b$$

where A is the transverse acceleration of τ.

Proof: As in the proof of Theorem 4.2.4 we let $H(u, \mathrm{v}) = [-g(\tau_u, \tau_u)]^{1/2}$ and observe that

$$L''(0) = \int_a^b \frac{\partial^2 H}{\partial \mathrm{v}^2}\Big|_{\mathrm{v}=0} du \ .$$

We have already shown that

$$\frac{\partial H}{\partial \mathrm{v}} = -\frac{g(\tau_u, \tau_{u\mathrm{v}})}{H}$$

Thus,

$$\frac{\partial^2 H}{\partial \mathrm{v}^2} = -\frac{1}{H^2}\left[H\frac{\partial}{\partial \mathrm{v}} g(\tau_u, \tau_{u\mathrm{v}}) - g(\tau_u, \tau_{u\mathrm{v}})\frac{\partial H}{\partial \mathrm{v}}\right]$$

$$= -\frac{1}{H}\left[g(\tau_{u\mathrm{v}}, \tau_{u\mathrm{v}}) + g(\tau_u, \tau_{u\mathrm{vv}}) + \frac{1}{H^2} g(\tau_u, \tau_{u\mathrm{v}})^2\right].$$

But, $\tau_{u\mathrm{v}} = \tau_{\mathrm{v}u}$ and $\tau_{u\mathrm{vv}} = \tau_{\mathrm{v}u\mathrm{v}} = \tau_{\mathrm{vv}u} + R_{\tau_\mathrm{v}\tau_u}\,\tau_\mathrm{v}$ so

$$\frac{\partial^2 H}{\partial \mathrm{v}^2} = -\frac{1}{H}\Big[g(\tau_{\mathrm{v}u}, \tau_{\mathrm{v}u}) + g(\tau_u, R_{\tau_\mathrm{v}\tau_u}\,\tau_\mathrm{v}) + g(\tau_u, \tau_{\mathrm{vv}u})$$

$$+ \frac{1}{H^2} g(\tau_u, \tau_{\mathrm{v}u})^2\Big].$$

Set $\mathrm{v} = 0$ to get $H = 1$, $\tau_u = \gamma'$, $\tau_\mathrm{v} = V$, $\tau_{\mathrm{v}u} = V'$, $\tau_{\mathrm{vv}} = A$ and $\tau_{\mathrm{vv}u} = A'$ and therefore

$$\frac{\partial^2 H}{\partial \mathrm{v}}\Big|_{\mathrm{v}=0} = -\left[g(V', V') + g(R_{V\gamma'}V, \gamma') + g(\gamma', A') + g(\gamma', V')^2\right].$$

From the fact that $R^b_{acd} = -R^a_{bcd}$ we conclude that $g(R_{V\gamma'}\, V, \gamma') = -g(R_{V\gamma'}\,\gamma', V)$. Moreover, since γ is a geodesic, $D_{\gamma'}\gamma' = 0$ so that differentiating $g(V, \gamma') = 0$ with respect to u gives $g(V', \gamma) = 0$. In addition, $\frac{d}{du} g(\gamma', A) = g(\gamma', A')$. Thus, we find that

$$\frac{\partial^2 H}{\partial \mathrm{v}^2}\Big|_{\mathrm{v}=0} = g(R_{V\gamma'}\,\gamma', V) - g(V', V') - \frac{d}{du} g(\gamma', A)\ .$$

Now integrate from a to b to obtain the required formula. Q.E.D.

We conclude with another bit of notation we shall use in the proof of Hawking's Theorem. Let us consider a globally hyperbolic spacetime M with Cauchy surface S and a future-directed timelike geodesic $\gamma: [-a, 0] \rightarrow M$, $a > 0$, which satisfies $g(\gamma', \gamma') = -1$, $\gamma(0) \in S$ and $g(\gamma'(0), \mathrm{v}) = 0$ for all v in $T_{\gamma(0)}(S)$. Denote by $V_0^{\perp}(\gamma)$ the vector space of all smooth vector fields V along γ which satisfy $g(V(u), \gamma'(u)) = 0$ for each u in $[-a, 0]$ and $V(-a) = 0$. Every member V of $V_0^{\perp}(\gamma)$ obviously has the property that $V(u)$ is spacelike or zero for each u and, moreover, $V(0) \in T_{\gamma(0)}(S)$. The elements of $V_0^{\perp}(\gamma)$ are the variation vector fields for orthogonal variations of γ which connect $\gamma(-a)$ with S. More precisely, a variation $\tau : [-a, 0] \times (-\delta, \delta) \rightarrow M$ of γ is said to *induce* the member V of $V_0^{\perp}(\gamma)$ if its longitudinal curves are future-directed and timelike, begin at $\gamma(-a)$, end at a point on S and if the variation vector field of τ is V. Any element of $V_0^{\perp}(\gamma)$ is induced by some such variation of γ (either construct your own or have a look at Lemma 49, page 297, of [O2]).

4.3. Covariant Differentiation in Submanifolds

The last section concluded with the observation that every element of $V_0^{\perp}(\gamma)$ is induced by a variation of γ. Notice that the final transverse curve in such a variation is contained in the hypersurface S. Since S is itself a manifold with metric (the restriction of M's metric to each $T_p(S) \subseteq T_p(M)$) a vector field along such a curve that is everywhere "tangent to S" will have covariant derivatives along the curve in both M in S. Of course, these two covariant derivatives will not be the same in general (consider, for example, a great circle on $S^2 \subseteq \mathbb{R}^3$). The last bit of preparation we require for the proof of Hawking's Theorem is a discussion of the famous "Gauss-Weingarten equations" which express several important relationships between these two derivatives.

Let us set the stage. We restrict our attention to a spacetime M, one of its Cauchy surface S and a point p in S. Then $T_p(S)$ is a 3-dimensional linear subspace of $T_p(M)$ all the nonzero elements of which are spacelike. Consequently, we can write

$$T_p(M) = T_p(S) \oplus T_p(S)^{\perp}$$

where $T_p(S)^{\perp}$ is the orthogonal complement of $T_p(S)$ in $T_p(M)$. $T_p(S)^{\perp}$ is therefore a one-dimensional linear subspace of $T_p(M)$ spanned by a timelike vector. We denote by $P : T_p(M) \rightarrow T_p(S)$ and $P^{\perp} : T_p(M) \rightarrow T_p(S)^{\perp}$ the projection maps. Now fix a v in $T_p(S)$ and let W be a smooth vector field of M that is *tangent to* S, i.e., that has the property that $W(q) \in T_q(S)$ for every q in S. We denote the covariant derivative in M of W in the direction v as usual by $D_{\mathrm{v}} W$. Thought of as a vector field on S, W also has covariant derivative in S in the direction v and this we denote $D_{\mathrm{v}}^S W$. The relationship between these two derivatives couldn't be simpler:

$$D_{\mathrm{v}}^S W = P(D_{\mathrm{v}} W)\ . \tag{10}$$

The proof of (10) is not difficult, but since we will be concerned primarily with the normal rather than the tangential component of $D_{\mathrm{v}} W$ we simply refer the reader to Theorem 1 in Chapter 1, Volume 3, of [Sp2]. To get at this normal component we introduce a bit of notation. Define a map

$$s : T_p(S) \times T_p(S) \to T_p(S)^{\perp}$$

as follows: For any v and w in $T_p(S)$ we set

$$s(\mathrm{v}, w) = P^{\perp}(D_{\mathrm{v}} W)\ , \tag{11}$$

where W is any smooth vector field on a neighborhood of p in M that is tangent to S on this neighborhood and extends w (i.e., $W(p) = w$). Of course, we must verify that this definition does not depend on the choice of the extension W of w.

Lemma 4.3.1. Let v and w be in $T_p(S)$. If W_1 and W_2 are two smooth vector fields on a neighborhood of p in M that are tangent to S and satisfy $W_1(p) = W_2(p) = w$, then $P^{\perp}(D_{\mathrm{v}} W_1) = P^{\perp}(D_{\mathrm{v}} W_2)$.

Proof. We define an operator K which takes a smooth vector field W defined on some neighborhood of p in M to the tangent vector

$$K(W) = P^{\perp}(D_{\mathrm{v}} W)\ .$$

Observe that, by Exercise 3.7.2, K is linear in W, i.e., $K(c_1 W_1 + c_2 W_2) = c_1 K(W_1) + c_2 K(W_2)$. Also observe that *if W is tangent to S* and f is any smooth real-valued function on some neighborhood of p in M, then

$$K(fW) = f(p) K(W) \qquad (W \text{ tangent to } S)$$

because $K(fW) = P^{\perp}(D_{\mathrm{v}}(fW)) = P^{\perp}(\mathrm{v}[f]W(p) + f(p) D_{\mathrm{v}} W) = P^{\perp}(f(p) D_{\mathrm{v}} W) = f(p) P^{\perp}(D_{\mathrm{v}} W) = f(p) K(W)$.

By linearity it will suffice to show that if W is tangent to S and $W(p) = 0$, then $K(W) = 0$. Choose a coordinate patch χ at p in M such that χ_1, χ_2 and χ_3 are tangent to S on some neighborhood U of p in M. Then $W = f^i \chi_i$, where f^i, $i = 1,2,3,4$, are C^{∞} real-valued functions on a neighborhood of p in M, $f^1(p)=f^2(p)=f^3(p)=0$ and, moreover, $f^4=0$ everywhere on $S \cap U$. It follows that $K(f^4\chi_4)=0$ and therefore $K(W)=K(f^1\chi_1)+K(f^2\chi_2)+K(f^3\chi_3)=f^1(p)K(\chi_1)+f^2(p)K(\chi_2)+f^3(p)K(\chi_3)=0.$ Q.E.D.

Consequently, $s(\mathrm{v},w)$ is well-defined. We now let N denote the unit normal field to S defined in section 3.7.

Exercise 4.3.1. Show that if v is in $T_p(S)$, then $D_\mathrm{v}N$ is in $T_p(S)$ also. *Hint.* N is a *unit* vector field.

The following result is true in a much more general context, but we shall record only that special case that is of interest to us (see Chapter 1, Volume 3 of [Sp2]).

Theorem 4.3.2. Let M be a spacetime, S a Cauchy surface in M, N the corresponding unit normal field, p a point in S and $\mathrm{v} \in T_p(S)$. Then for any smooth vector field W on a neighborhood of p in M that is tangent to S we have

$$D_\mathrm{v}W = D_\mathrm{v}^S W + s(\mathrm{v}, W(p)) \qquad \text{(the \textit{Gauss formula})} \qquad (12)$$

and

$$g(D_\mathrm{v}N, W) = -g(N, D_\mathrm{v}W) = -g(N, s(\mathrm{v}, W(p))) \quad \text{(the \textit{Weingarten equations})}\ . \quad (13)$$

Proof: The Gauss formula is just a restatement of (10) and (11). To prove the equalities in (13) we begin with $g(N,W)=0$ on S to obtain $0=\mathrm{v}[g(N,W)]=g(N,D_\mathrm{v}W)+g(D_\mathrm{v}N,W)$ which gives the first equality. The second is just the definition of $s(\mathrm{v},W(p))$ as the normal component of $D_\mathrm{v}W$. Q.E.D.

4.4. Proof of the Singularity Theorem

For convenience we shall restate our major result (Theorem 3.8.1).

Hawking's Theorem: Let M be a (time orientable, stably causal, globally hyperbolic) spacetime which satisfies

1. Ric $(V,V) \geq 0$ for all timelike tangent vectors V, and
2. There exists a Cauchy surface S in M on which the mean curvature is bounded below by some positive constant k, i.e., $H_S(p) \geq k$ for each p in S.

Then M is timelike geodesically incomplete. More precisely, if $\mu : (-u_0, 0] \to M$ is any future-directed timelike geodesic such that $g(\mu', \mu') = -1$, $\mu(0) \in S$ and $\mu'(0)$ is normal to S, then $-u_0 \geq -3/k$.

Proof. Let $-v_0 \in (-u_0, 0]$ be arbitrary. We will show that $-v_0 \geq -3/k$ and conclude therefore that $-u_0 \geq -3/k$. Let $p = \mu(-v_0)$.

Figure 4.4

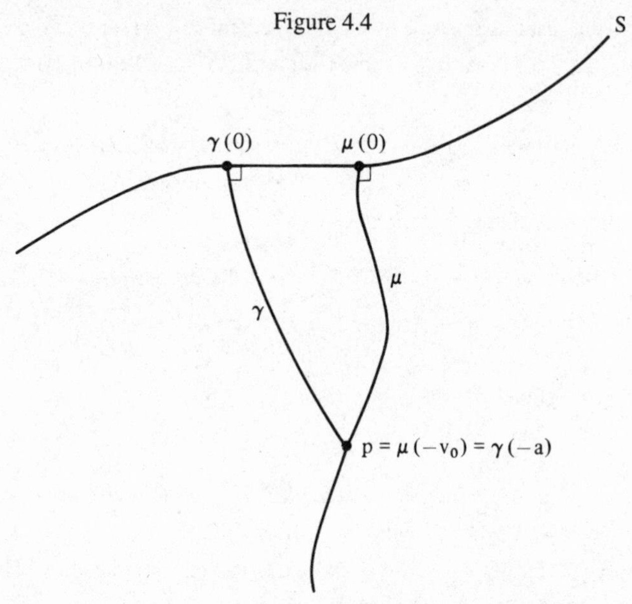

By Exercise 4.1.1 there exists a future-directed timelike geodesic $\gamma : [-a, 0] \to M$ $(a > 0)$ such that $\gamma(-a) = p$, $\gamma(0) \in S$, $\gamma'(0)$ is orthogonal to S, $g(\gamma', \gamma') = -1$ and such that the proper time length a of γ is greater than or equal to the proper time length of any other smooth timelike curve in M from p to a point in S. Since μ and γ are both unit speed geodesics they are both parametrized by proper time so, in particular, the maximality of γ implies that $v_0 \leq a$ and it will therefore suffice to show that

$$-a \geq -3/k \ . \tag{14}$$

Select an orthonormal basis $\{W_1(-a), W_2(-a), W_3(-a), \gamma'(-a)\}$ at p and let $\{W_1(u), W_2(u), W_3(u), \gamma'(u)\}$ be the vector fields along γ obtained by parallel translation (see section 3.3 and observe that since γ is a geodesic its velocity vector γ' *is* parallel along γ). By Theorem 3.3.4 these vectors form an orthonormal basis for $T_{\gamma(u)}(M)$ for each u in $[-a, 0]$. Moreover, the W_i satisfy all the requirements for elements of $V_0^\perp(\gamma)$ except that they do not vanish at $u = -a$. To obtain elements of $V_0^\perp(\gamma)$ we define vector fields V_i along γ by

$$V_i(u) = \frac{a+u}{a} W_i, \; i = 1,2,3,$$

for each u in $[-a, 0]$. Each V_i is in $V_0^\perp(\gamma)$. Consequently, each of the V_i is induced by a variation τ_i of γ and each τ_i in turn gives rise to a length function L_i: $(-\delta_i, \delta_i) \to \mathbb{R}$. By maximality of γ each of these length functions must satisfy $L_i''(0) \leq 0$ so $\sum_{i=1}^{3} L_i''(0) \leq 0$.

According to Theorem 4.2.4 we therefore have

$$0 \leq \sum_{i=1}^{3} \left[\int_{-a}^{0} \{g(V'_i, V'_i) - g(R_{V_i\gamma'}\gamma', V_i)\}\, du + g(A_i(0), \gamma'(0)) \right] \tag{15}$$

(note that $A_i(-a) = 0$ since the transverse curve at $-a$ is constant).

Exercise 4.4.1. Let $\alpha : I \to M$ be a smooth curve, $f(u)$ a smooth real-valued function on I and W a smooth vector field on α. Show that $D_{\alpha'}(fW) = f'W + fD_{\alpha'}W$.

Since each W_i is parallel along γ it follows from Exercise 4.4.1 that $V'_i = \frac{1}{a} W_i$. Consequently, $g(V'_i, V'_i) = \frac{1}{a^2} g(W_i, W_i) = 1/a^2$ so $\int_{-a}^{0} g(V'_i, V'_i)\, du = 1/a$ and therefore

$$\sum_{i=1}^{3} \int_{-a}^{0} g(V'_i, V'_i)\, du = \frac{3}{a} \; . \tag{16}$$

To handle the second term in (15) we observe first that from the definition of the curvature operator it is clear that $R_{V_i\gamma'}\gamma' = \frac{a+u}{a} R_{W_i\gamma'}\gamma'$ so that $\sum_{i=1}^{3} g(R_{V_i\gamma'}\gamma', V_i) = (\frac{a+u}{a})^2 \sum_{i=1}^{3} g(R_{W_i\gamma'}\gamma', W_i)$. But we claim that this last sum is actually $\text{Ric}(\gamma', \gamma')$. To see this we observe that the obvious symmetry $R^a_{bcd} = -R^a_{bdc}$ of the Riemann tensor implies that $R_{XY}Z = -R_{YX}Z$ so, in particular, $R_{XX}X = 0$. Consequently, our assertion follows from Exercise 4.2.3. Thus,

$$\sum_{i=1}^{3} \int_{-a}^{0} g(R_{V_i\gamma'}\gamma', V_i)\, du = \int_{-a}^{0} (\frac{a+u}{a})^2 \,\text{Ric}(\gamma', \gamma')\, du \; . \tag{17}$$

The last term in (15) is $\sum_{i=1}^{3} g(A_i(0), \gamma'(0))$. Since $\gamma'(0)$ is a future-directed unit timelike normal to S it must coincide with $N(\gamma(0))$, where N is the unit normal field defined in section 3.7. Now, $\{W_1(0), W_2(0), W_3(0), N(\gamma(0))\}$ is an orthnormal basis for $T_{\gamma(0)}(M)$ with $\{W_1(0), W_2(0), W_3(0)\}$ spanning $T_{\gamma(0)}(S)$. N is a smooth vector field on all of M. Extend $\{W_1(0), W_2(0), W_3(0)\}$ to vector fields $\{\overline{W}_1, \overline{W}_2, \overline{W}_3\}$ on some neighborhood U of $\gamma(0)$ in M such that, for each p in $U \cap S$, $\{\overline{W}_1(p), \overline{W}_2(p), \overline{W}_3(p)\}$ is an orthonormal basis for $T_p(S)$. Now we compute

$$\begin{aligned}
\sum_{i=1}^{3} g(D_{\overline{W}_i}\overline{W}_i, N)(\gamma(0)) &= \sum_{i=1}^{3} g(D_{\overline{W}_i(\gamma(0))}\overline{W}_i, N(\gamma(0))) \\
&= \sum_{i=1}^{3} g(D_{V_i(0)}\overline{W}_i, N(\gamma(0))) \\
&= \sum_{i=1}^{3} g(D^{S}_{V_i(0)}\overline{W}_i + s(V_i(0), \overline{W}_i(0)), N(\gamma(0))) \\
&= \sum_{i=1}^{3} g(s(V_i(0), V_i(0)), N(\gamma(0)))
\end{aligned}$$

Exercise 4.4.2. Show that the last sum above is equal to $\sum_{i=1}^{3} g(A_i(0), N(\gamma(0)))$ and conclude that

$$\sum_{i=1}^{3} g(A_i(0), \gamma'(0)) = \sum_{i=1}^{3} g(D_{\overline{W}_i}\overline{W}_i, N)(\gamma(0)) \,. \tag{18}$$

On the other hand, we claim that the right-hand side of (18) is just $-H_S(\gamma(0))$. To see this we write

$$\begin{aligned}
\sum_{i=1}^{3} g(D_{\overline{W}_i}\overline{W}_i, N)(\gamma(0)) &= \sum_{i=1}^{3} g(D_{\overline{W}_i(\gamma(0))}\overline{W}_i, N(\gamma(0))) \\
&= -\sum_{i=1}^{3} g(D_{\overline{W}_i(\gamma(0))}N, \overline{W}_i(\gamma(0))) \text{ by } (13) \\
&= -\operatorname{div} N(\gamma(0))
\end{aligned}$$

by Exercise 3.7.5 and the fact that $g(D_{N(\gamma(0))}N, N(\gamma(0))) = 0$ by the Hint in Exercise 4.3.1. But $\operatorname{div} N(\gamma(0)) = H_S(\gamma(0))$ by definition so we conclude that

$$\sum_{i=1}^{3} g(A_i(0),\ \gamma'(0)) = -H_S(\gamma(0)) \ . \tag{19}$$

To conclude the proof we substitute (16), (17) and (18) into (15) and obtain

$$0 \leq \frac{3}{a} - \int_{-a}^{0} \left(\frac{a+u}{a}\right)^2 \mathrm{Ric}\,(\gamma', \gamma')\, du - H_S(\gamma(0)) \ . \tag{20}$$

Since γ' is timelike assumption #1 of the theorem implies that the integral in (20) is non-negative so that

$$0 \leq \frac{3}{a} - H_S(\gamma(0)) \ . \tag{21}$$

Moreover, since $\gamma(0)$ is the S, assumption #2 gives $H_S(\gamma(0)) \geq k$ so that

$$0 \leq \frac{3}{a} - k\,, \tag{22}$$

which is equivalent to (14) Q.E.D.

4.5. The Existence of Geodesics

And so we have, at last, a proof of Hawking's Theorem. Well, not quite. We still must deal with the as yet unproved Theorem 4.1.1 and this is no simple matter. Indeed, the proof of this crucial lemma is long, difficult and at times rather technical. In this final section we shall attempt to provide something of the flavor of the argument while at the same time evading most of the technicalities. We sketch a proof based on a number of preliminary lemmas. For each of these lemmas we shall be content with either a sketch of the proof or simply a reference where the proof can be found. Even this requires some work however and we must begin with a few definitions.

If M is a spacetime and p is in M, then a geodesically convex normal neighborhood N of p is called a *simple region* if its closure in M is compact and contained in another geodesically convex normal neighborhood of p. Every p in M has a local base consisting entirely of simple regions. We denote by U_p the maximal open subset of $T_p(M)$ on which $\exp_p : U_p \to N$ is a diffeomorphism onto N. A map $\Omega : N \times N \to \mathbb{R}$ called the *world function* (in [Pen] and [Sy2]) or the *geometric energy function* (in [SW2]) is defined by

$$\Omega(p,q) = g(\exp_p^{-1}(q),\ \exp_p^{-1}(q)) \ .$$

Thus, $\Omega(p,q)$ is the squared Lorentzian length of the unique geodesic segment $\mu : [0,1] \to N$ such that $\mu'(0) = \exp_p^{-1}(q)$. Indeed,

$$\int_0^1 g(\mu'(t), \mu'(t))\, dt = \int_0^1 g(\mu'(0), \mu'(0))\, dt$$

$$= g(\mu'(0), \mu'(0)) \int_0^1 dt$$

$$= g(\exp_p^{-1}(q), \exp_p^{-1}(q)) .$$

Observe that $\Omega(p,q) = \Omega(q,p)$ and that $\Omega(p,q)$ is > 0, < 0 or $= 0$ according as μ as spacelike, timelike or null. Moreover Ω is C^∞ on the product manifold $N \times N$. For each fixed p in N we define $\Omega_p : N \to \mathbb{R}$ by $\Omega_p(q) = \Omega(p,q)$. For each k in $\mathbb{R}$ we let

$$H_{p,k} = \{q \in N : q \neq p \text{ and } \Omega_p(q) = k\}$$

and observe that $H_{p,k}$ is a smooth 3-dimensional submanifold of N (and therefore of M). This is most easily seen by choosing Minkowski normal coordinates for N and using the fact that $\exp_p : U_p \to N$ is a diffeomorphism. We denote by $C_T^+(p) \subseteq T_p(M)$ the open solid cone of future-directed timelike vectors in $T_p(M)$; $C_T^-(p)$ is defined dually by replacing "future" with "past". Finally, we set

$$N_T^{\pm}(p) = \exp_p(C_T^{\pm}(p) \cap U_p) \subseteq N$$

and observe that $N_T^+(p)$ and $N_T^-(p)$ are open and disjoint and that, moreover, $q \in N_T^+(p) \cup N_T^-(p)$ if and only if μ is timelike.

We show next that inside a simple region radial geodesics from p are orthogonal to level hypersurfaces of Ω_p, i.e., we establish

Lemma 4.5.1 (*The Gauss Lemma*). Let N be a simple region containing p, k a real number and q a point in $H_{p,k}$. Then the unique geodesic $\mu : [0,1] \to N$ with $\mu'(0) = \exp_p^{-1}(q)$ is orthogonal to $H_{p,k}$ at q.

Exercise 4.5.1. Let $\tau : [a,b] \times (-\delta,\delta) \to M$ be a variation of the curve $\mu(u) = \tau(u, 0)$ and assume that each longitudinal curve $u \to \tau(u, v_0)$ is a geodesic with $g(\tau_u, \tau_u)$ the same constant value on each. Show that $g(\tau_u, \tau_v)$ is constant along μ. **Hint.** Calculate $\frac{d}{du} g(\tau_u, \tau_v)$ from the product rule and use $\tau_{vu} = \tau_{uv}$.

The Gauss Lemma follows easily from Exercise 4.5.1 in the following way: Let $\alpha : (-\delta,\delta) \to H_{p,k}$ be an arbitrary smooth curve in $H_{p,k}$ with $\alpha(0) = q$. We must show that $g(\alpha'(0), \mu'(1)) = 0$.

Figure 4.5

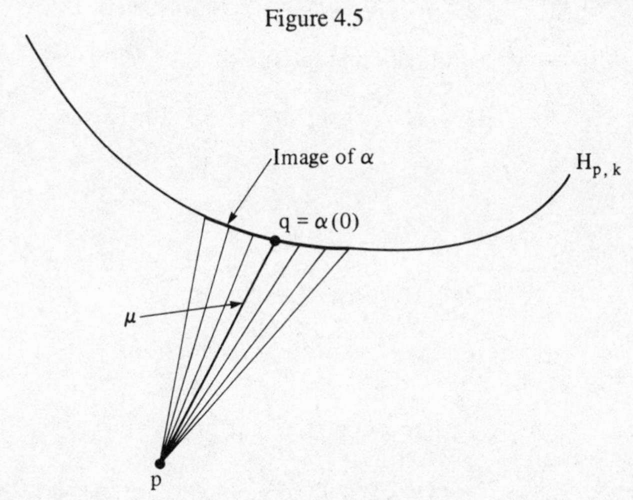

Intuitively, the idea is that by allowing q to vary along α the corresponding geodesics through p generate a variation of μ which satisfies the hypotheses of Exercise 4.5.1, is constant at p and so is orthogonal to $\mu'(0)$. By the Exercise this orthogonality propagates to q along μ.

Exercise 4.5.2. Fill in the details. **Hint.** Define $\tau : [0,1] \times (-\delta,\delta) \to M$ by $\tau(u, v) = \exp_p(u \cdot \exp_p^{-1}(\alpha(v)))$ and show that $g(\tau_v(1,0), \tau_u(1,0)) = 0$.

We have already observed (section 3.3) that in $\mathcal{M}$ the gradient of the squared length function is "radial". The corresponding result for an arbitrary simple region is

Lemma 4.5.2. Let N be a simple region containing p, q in $N - \{p\}$ and $\mu : [0,1] \to N$ the unique geodesic in N with $\mu'(0) = \exp_p^{-1}(q)$. Then

$$\nabla \Omega_p(q) = 2\mu'(1) \ . \tag{23}$$

This is proved by first observing that $\nabla\Omega_p(q) = 0$ only if $q = p$ so that $\nabla\Omega_p$ is nonzero on $N - \{p\}$. Setting $k = \Omega_p(q)$ we have $q \in H_{p,k}$. By the Gauss Lemma, $\mu'(1)$ is a nonzero normal vector to $H_{p,k}$. But, by Lemma 3.5.13, $\nabla\Omega_p(q)$ is also a nonzero normal vector to $H_{p,k}$. Since $T_q(H_{p,k})$ is 3-dimensional and $T_q(M)$ is 4-dimensional we find that $\nabla\Omega_p(q)$ and $\mu'(1)$ must be linearly dependent, i.e.,

$$\nabla\Omega_p(q) = a\mu'(1) \ ,$$

where a is a nonzero real number which could *a priori* depend (continuously) on q. We show that, in fact, $a = 2$ for any q. First we compute, for any t in $[0,1]$,

$$\begin{aligned}\Omega_p\,(\mu(t)) &= g(\exp_p^{-1}(\mu(t)), \exp_p^{-1}(\mu(t))) \\ &= g(t \bullet \exp_p^{-1}(q), t \bullet \exp_p^{-1}(q)) \qquad \text{(by Exercise 3.27)} \\ &= t^2 k \ .\end{aligned}$$

Thus,

$$\frac{d}{dt}\Omega_p\,(\mu(t)) = 2kt, \ 0 \le t \le 1 \ .$$

But, by Lemma 3.3.2,

$$2kt = g\,(\nabla\Omega_p\,(\mu(t)), \mu'(t)) \ .$$

Evaluating at $t = 1$ gives $2k = g\,(\nabla\Omega_p(q), \mu'(1)) = g\,(a\mu'(1), \mu'(1)) = ak$ so $2k = ak$. Now, if $k \neq 0$, i.e., if μ is not a null geodesic, this implies $a = 2$. On the other hand, if μ is null, $a = 2$ by continuity. In any case, (23) is proved.

From this one obtains the long-promised generalization of Theorem 1.5.6.

Lemma 4.5.3. Let N be a simple region with p and q in N. Suppose there exists a smooth, future-directed timelike curve (equivalently, trip) $\alpha : [0,1] \to N$ with $\alpha(0) = p$ and $\alpha(1) = q$. Then the unique geodesic segment in N from p to q must also be timelike and future-directed.

The proof goes something like this: We let $\beta : [0,1] \to T_p(M)$ be the curve in $T_p(M)$ obtained by "lifting" α, i.e.,

$$\beta = \exp_p^{-1} \circ \alpha \ .$$

Identifying the velocity vector to β at 0 with an element of $T_p(M)$ in the usual way we have $\beta'(0) = \alpha'(0)$ so $\beta'(0)$ is inside the upper time cone $C_T^+(p)$ at 0 in $T_p(M)$. Since $C_T^+(p)$ is open and convex in $T_p(M)$, β must at least initially remain in $C_T^+(p)$, i.e., there exists an $\varepsilon > 0$ such that $\beta[0,\varepsilon] \subseteq C_T^+(p)$. Thus, $\alpha[0,\varepsilon] \subseteq N_T^+(p)$. Now we show that, having once entered $N_T^+(p)$, α cannot escape, i.e., that $\alpha(1)$ is also in $N_T^+(p)$. To see this we suppose $\alpha(1) \notin N_T^+(p)$ and argue by contradiction. If $\alpha(1) \notin N_T^+(p)$, then $\beta(1) \notin C_T^+(p)$. In order to exit $C_T^+(p)$, β must either pass through 0 in $T_p(M)$ or encounter the upper null cone in $T_p(M)$ so there exists a t_0 in $(\varepsilon, 1]$ at which $\Omega_p(\alpha(t_0)) = 0$. Moreover, $\beta(t_0) \neq 0$ since then $\alpha(t_0) = \exp_p(0) = p = \alpha(0)$, but M contains no closed timelike curves (stable causality). Thus, $\beta(t_0)$ is on the upper null cone in $T_p(M)$ so that if $q = \alpha(t_0)$ and $\mu : [0,1] \to N$ is the unique geodesic segment in N from p to q, then $\mu'(1)$ is null and future-directed. By Lemma 4.5.2, $\nabla \Omega_p(q)$ is also future null. Since $\alpha'(t_0)$ is future timelike, Theorem 1.3.1 implies that $g(\nabla\Omega_p(\alpha(t_0)), \alpha'(t_0)) < 0$ so that

$$\frac{d}{dt}\Omega_p(\alpha(t))\bigg|_{t=t_0} < 0 .$$

i.e., $\Omega_p(\alpha(t))$ is decreasing on some interval about t_0. Since $\Omega_p(\alpha(t_0)) = 0$, $\Omega_p(\alpha(t))$ must be positive immediately to the left of t_0. We find then that whenever α hits the upper boundary of $N_T^+(p)$, i.e., whenever β hits the upper null cone at 0 in $T_p(M)$, $\Omega_p(\alpha(t))$ must be positive immediately to the left. Now consider the set $\{t \in [0,1] : \alpha(t) \in N_T^+(p)\}$. It is nonempty and bounded above so its supremum t_1 exists. However, t_1 must obviously be on the upper boundary of $N_T^+(p)$ so $\Omega_p(\alpha(t))$ is positive immediately to the left of t_1. But then $\alpha(t)$ cannot be in $N_T^+(p)$ immediately to the left of t_1 and this contradicts the definition of t_1.

These preliminary results will now be used to exhibit a vital characteristic of globally hyperbolic spacetimes, indeed, one that is often taken as their defining characteristic. Again, we must begin with some definitions. A smooth curve $\alpha : I \to M$ is said to be *causal* if its velocity vector $\alpha'(t)$ is either timelike or null for each t in I. We shall use the term *causal geodesic* for a geodesic which is either timelike or null (*possibly degenerate*). A *causal trip* is defined in the same way as a trip except that causal geodesics replace timelike geodesics. Lemma 4.5.3 easily generalizes to the case in which there is a causal trip from p to q that is not a null geodesic.

Lemma 4.5.4. Let N be a simple region with p and q in N. Suppose there exists a smooth, future-directed causal curve (equivalently, causal trip) from p to q that is not a null geodesic. Then the unique geodesic segment in N from p to q is timelike and future-directed.

We say that p *causally precedes* q and write $p \leq q$ if and only if there exists a causal trip (equivalently, causal curve, possibly degenerate) from p to q. The *causal future* of any p in M is denoted $J^+(p)$ and defined by $J^+(p) = \{q \in M : p \leq q\}$. $J^-(p)$ is defined dually and for any $S \subseteq M$, $J^\pm(S) = \bigcup_{p \in S} J^\pm(p)$.

Exercise 4.5.3. Describe $J^{\pm}(p)$ for an arbitrary p in $\mathcal{M}$, $\mathcal{D}$ and $\mathcal{E}$.

One routine fact we shall require is the following "transitivity" relation between $\leq$ and $\ll$:

$$p \leq q \quad \text{and} \quad q \ll r \ \text{ implies } \ p \ll r\ . \tag{24}$$

Similarly,

$$p \ll q \quad \text{and} \quad q \leq r \ \text{ implies } \ p \ll r\ . \tag{25}$$

These are, of course, obvious in $\mathcal{M}$, $\mathcal{D}$ and $\mathcal{E}$, but true in any spacetime. Either construct your own proof or consult Proposition 2.18 of [Pen].

Exercise 4.5.4. Prove each of the following:

(a) For any p in M the closure of $I^+(p)$ in M is $\overline{I^+(p)} = \{q \in M : I^+(q) \subseteq I^+(p)\}$.

(b) $p \leq q$ implies $I^+(q) \subseteq I^+(p)$.

(c) $\overline{J^+(p)} = \overline{I^+(p)}$.

(d) $J^+(p)$ need not be closed. **Hint.** Your example will have to be in a spacetime that is *not* globally hyperbolic.

Our first major step toward Theorem 4.1.1. asserts in effect that a globally hyperbolic spacetime is "causally compact" between any Cauchy surface S and any p not in S.

Lemma 4.5.5. Let S be any Cauchy surface in M. Then for any p in the interior of $D^+(S)$, $J^-(p) \cap J^+(S)$ is compact. Similarly, for any $p \in \operatorname{int} D^-(S)$, $J^+(p) \cap J^-(S)$ is compact.

Remark. For a Cauchy surface S, the boundaries of $D^+(S)$ and $D^-(S)$ are both S so any $p \notin S$ is in the interior of one of these sets.

One argues by contradiction. Thus, we assume $p \in \operatorname{int} D^+(S)$, but nevertheless $K = J^-(p) \cap J^+(S)$ is not compact. We shall indicate how from this assumption one can actually build a past inextendible trip through some point of $D^+(S)$ which fails to meet S and this, of course, is a contradiction (here we must begin to rely on some of the theory of metric spaces).

Begin by observing that, even if not compact, K is paracompact and Lindelöf (i.e., for any open cover $\{U_\alpha\}$ of K there exists a countable open cover $\{V_1, V_2, V_3, \ldots\}$ with each V_i contained in some U_α and which is *locally finite*, i.e., has the property that each point in K has a neighborhood which intersects only finitely many of the V_i). Since the simple regions of M form a base for the open sets in M one can therefore construct a countable cover $\{N_i\}_{i=1}^{\infty}$ of K by simple regions in M which is locally finite, but has no finite subcover of K. For each $i = 1,2,3,...$ select a_i in $N_i \cap K$ with $a_j \neq a_i$ if $j \neq i$. By local finiteness the sequence $\{a_i\}_{i=1}^{\infty}$ has no accumulation point in K. We construct our trip inductively as follows: Let $x_0 = p$. Then x_0 is in some N_{i_0}. Choose y_0 in $N_{i_0} \cap I^+(x_0) \cap D^+(S)$ (remember that $x_0 \in \operatorname{int} D^+(S)$).

Figure 4.6

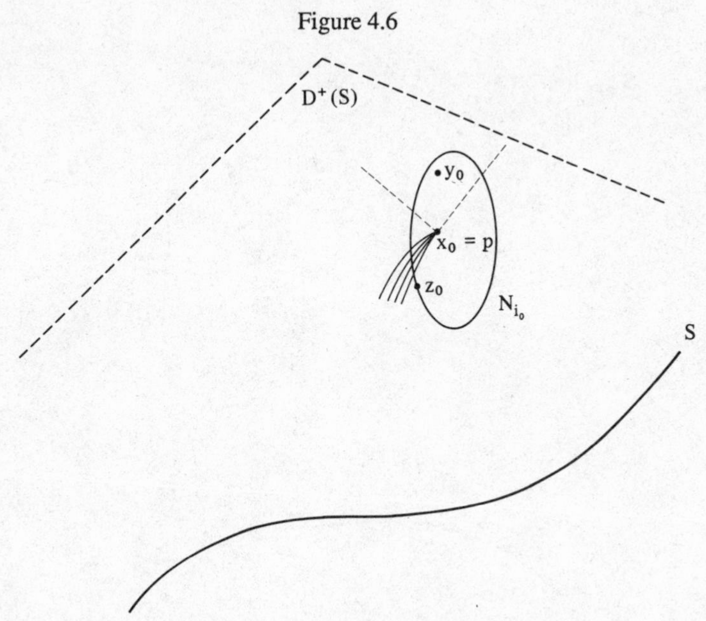

Since each a_i is in $J^-(x_0)$ there exist causal trips in M from each a_i to x_0. Since each $\overline{N}_i$ is compact, infinitely many of the a_i must lie outside of N_{i_0}. Thus, infinitely many of the causal trips from the a_i to x_0 must meet the boundary ∂N_{i_0} of N_{i_0}. Since ∂N_{i_0} is compact these intersection points must have an accumulation point z_0 in ∂N_{i_0}. Since $\overline{N}_{i_0}$ is contained in some larger geodesically convex normal neighborhood, continuity of the world function gives $z_0 \leq x_0$. Together with $x_0 \ll y_0$ this yields

$$z_0 \ll y_0 \ .$$

By Lemma 4.5.3 the unique geodesic in $\overline{N}_{i_0}$ joining z_0 and y_0 (which we shall here denote $z_0 y_0$) is future-timelike.

Now, z_0 is not in N_{i_0}. However, z_0 is in $J^-(x_0)$. Moreover, if z_0 were not in $J^+(S)$, then no past inextendible causal (in particular, timelike) trip from z_0 could hit S. One such trip together with z_0y_0 would be a past inextendible trip from $y_0 \in D^+(S)$ which misses S and we are done. Thus, we may assume z_0 is in $J^+(S)$ so $z_0 \in K$. Consequently, there is an N_{i_1} $(\neq N_{i_0})$ with $z_0 \in N_{i_1}$. Select x_1 and y_1 on the portion of z_0y_0 in N_{i_1} such that

$$z_0 \ll x_1 \ll y_1 \ll y_0 .$$

Figure 4.7

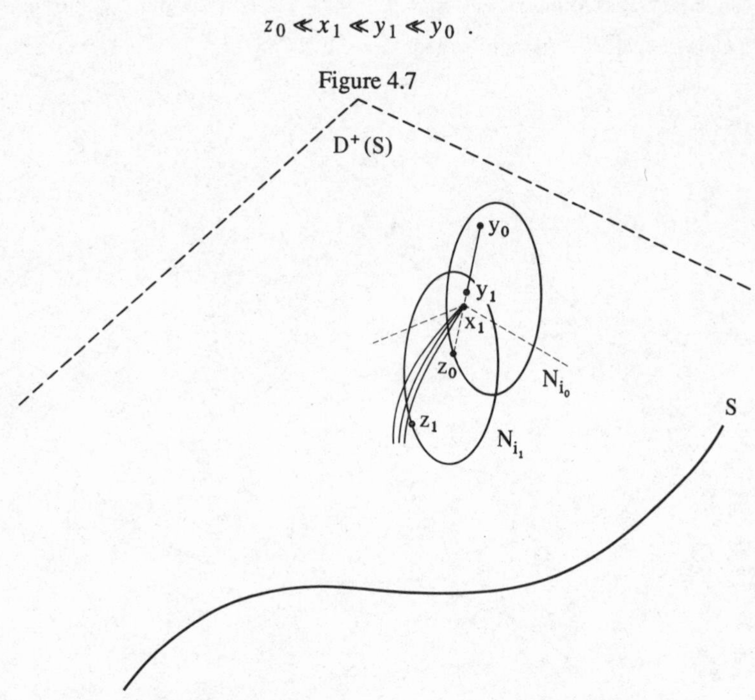

Now, $I^-(x_1)$ is an open neighborhood of z_0 (Lemma 3.3.10) so infinitely many of the causal trips from a_i to x_0 enter it. Select a point on each inside $I^-(x_1)$. Then $a_i \ll$ this point $\ll x_1$ implies $a_i \in I^-(x_1)$. If all but finitely many of these a_i were on a single causal trip to x_0 they would have to accumulate. Thus, infinitely many of the a_i must lie in $I^-(x_1)$. Infinitely many of these must lie outside N_{i_1} so again there exists a z_1 in ∂N_{i_1} which is an accumulation point of the set of intersections with ∂N_{i_1} of trips from a_i to x_1. Again we find that $z_1 \leq x_1$ so $z_1 \ll y_1$ and so the geodesic z_1y_1 is future-timelike. z_1 is not in N_{i_1} so there is an N_{i_2} with $z_1 \in N_{i_2}$. As before we select x_2 and y_2 on the portion of z_1y_1 in N_{i_2} such that

$$z_1 \ll x_2 \ll y_2 \ll y_1 .$$

Figure 4.8

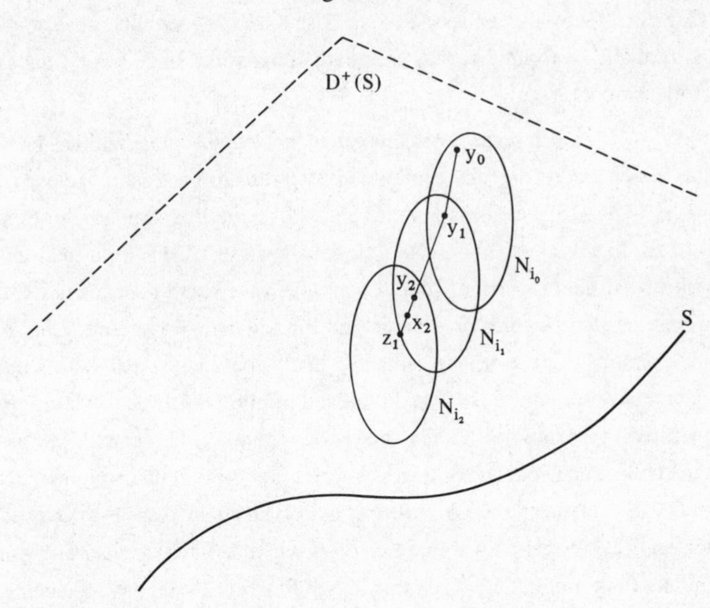

Continuing in this way we obtain a sequence $y_0, y_1, y_2, \ldots$ with

$$\ldots \ll y_2 \ll y_1 \ll y_0 \in D^+(S) \ .$$

We consider the trip whose segments are $\ldots, y_2 y_1, y_1 y_0$. This trip has future endpoint $y_0 \in D^+(S)$. It is past inextendible (i.e., has no past endpoint) since if q were a past endpoint it would be an accumulation point of the y_i's and this is inconsistent with the local finiteness of $\{N_i\}$. Finally, we show that this trip cannot intersect S by assuming that it does. Thus, suppose there is an s in S which is on the image of our trip.

Figure 4.9

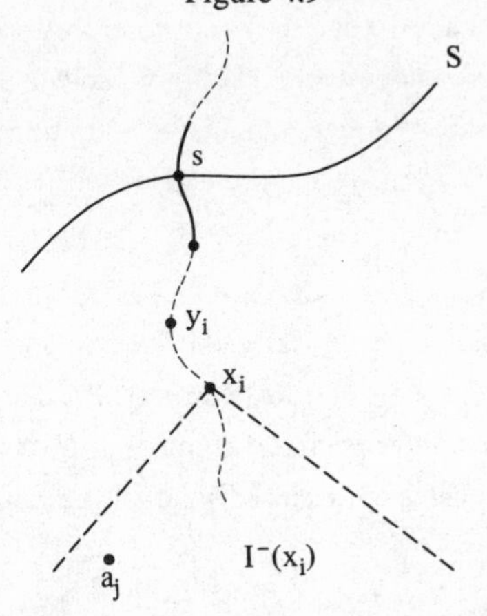

Then some y_i is in $I^-(s)$. But there exists an a_j such that $a_j \ll x_i \ll y_i$ so it follows that $a_j \ll s$. However, there also exists an s_0 in S such that $s_0 \le a_j$ (because $a_j \in J^+(S)$). Now, $s_0 \le a_j$ and $a_j \ll s$ imply $s_0 \ll s$ and this is impossible since S is a Cauchy surface and therefore achronal.

Theorem 4.1.1 is, in a sense we shall explain, analogous to the familiar theorem from the calculus which asserts that any continuous, real-valued function on a closed, bounded interval $[a,b]$ in $\mathbb{R}$ achieves a maximum value. Observe that similar assertions for the intervals $[a,b)$, $[a,\infty)$, $(-\infty,b]$, etc. are obviously false. Indeed, in real analysis it becomes clear that the property of $[a,b]$ upon which the result depends is its "compactness" and that the same theorem can be proved for any compact subset of any Euclidean space (see, e.g., [Sp1]). In fact, one can show that a continuous, real-valued function on any compact metric space achieves maximum and minimum values. This suggests a procedure for proving Theorem 4.1.1 that may (very roughly) be summarized as follows: Consider the collection C of all causal curves which join a point in S with the fixed point p in, say, $\operatorname{int} D^-(S)$. Define a metric on the set C in such a way that (1) the metric space C is compact, and (2) the real-valued function on C which assigns to every element of C its Lorentzian length is continuous. Then this length function would have to achieve a maximum value on some causal curve in C which we might then try to show is necessarily a geodesic (as it certainly is in, say, $\mathcal{M}$). Needless to say, this isn't quite as easy as it sounds. Indeed, we shall find that it is not possible to achieve all that we ask. There is no reasonable way to define a compact metric on the set of all smooth causal curves from p to S since limits of smooth curves (functions) are generally not smooth. This will necessitate extending the notion of "causal" to curves that are continuous, but not necessarily smooth. On this larger set of curves we shall find that there is a natural compact metric (compactness depends in a crucial way upon the compactness of $J^+(p) \cap J^-(S)$). Of course, we must also extend the notion of "Lorentzian length" to this new set of curves (for which a velocity vector need not exist) and hope that, having done so, the length function is continuous relative to our metric. Unfortunately, it turns out not to be continuous, but *is* upper semicontinuous so that, although it will not achieve a minimum in general, it will achieve a maximum as we require. We now set about filling in some of the details.

We shall say that a continuous curve $\alpha : (a,b) \to M$ is *future causal* if for each t_0 in (a,b) there exists an $\varepsilon > 0$ and a geodesically convex normal neighborhood $U(\alpha(t_0))$ of $\alpha(t_0)$ such that $\alpha(t_0 - \varepsilon, t_0 + \varepsilon) \subseteq U(\alpha(t_0))$ and such that for all t_1 and t_2 with $t_0 - \varepsilon < t_1 < t_2 < t_0 + \varepsilon$ the unique geodesic segment in $U(\alpha(t_0))$ from $\alpha(t_1)$ to $\alpha(t_2)$ is future causal. If α is defined on a closed interval we make the obvious one-sided modifications of the definition at each endpoint. Henceforth, "future causal" will refer to a continuous curve of this sort and we will explicitly specify "smooth" when it is intended.

Now we consider a Cauchy surface S in M and a point p in, say, int $D^-(S)$. We set

$$K = J^+(p) \cap J^-(S) \ ,$$

recall that K is compact (Lemma 4.5.5) and denote by $C_K(p,S)$ the set of all future causal curves in K from p to a point in S. Letting d denote the natural metric on M (i.e., the restriction to M of the usual metric on the ambient Euclidean space) we recall that for any $E \subseteq M$ and any point p the distance from p to E is defined by

$$d(p,E) = \inf \{d(p,e) : e \in E\} \ .$$

For each α in $C_K(p,S)$ and each $\varepsilon > 0$ we let

$$V_\varepsilon(\alpha) = \{y \in M : d(y, \operatorname{Im}(\alpha)) < \varepsilon\}$$

be the "ε-band" about α:

Figure 4.10

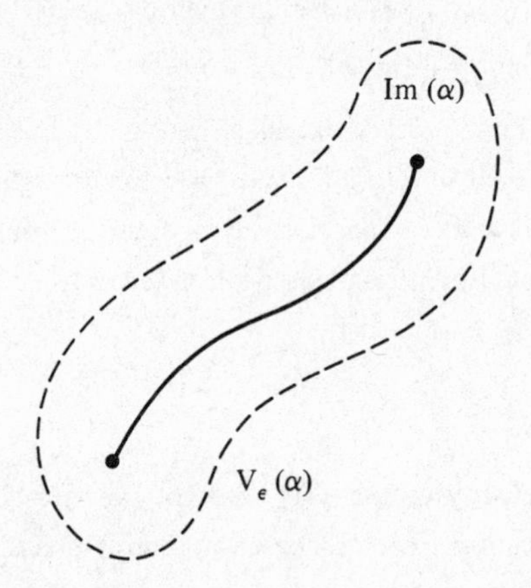

Finally, if α_1 and α_2 are both in $C_K(p,S)$ we define

$$\rho(\alpha_1,\alpha_2) = \inf \{\varepsilon : \varepsilon > 0,\ \alpha_1 \subseteq V_\varepsilon(\alpha_2),\ \alpha_2 \subseteq V_\varepsilon(\alpha_1)\} \ .$$

ρ is then the restriction to $C_K(p,S)$ (i.e., the restriction to the set of images of the α's in $C_K(p,S)$) of the Hausdorff metric on the closed subsets of M. Intuitively, $\rho\,(\alpha_1,\alpha_2)$ is the radius of the "smallest" band which when put around either curve encloses the other as well.

Exercise 4.5.5. Let $\alpha_1, \alpha_2, \ldots$ be elements of $C_K(p,S)$ and reparametrize each by an affine change of parameter so that they are all defined on $[0,1]$. The sequence $\{\alpha_n\}$ converges to α in the metric space $(C_K(p,S), \rho)$ if and only if $\rho(\alpha, \alpha_n) \to 0$ as $n \to \infty$. Show that this is the case if and only if $\{\alpha_n(0)\} \to \alpha(0)$, $\{\alpha_n(1)\} \to \alpha(1)$ and for each open set V containing Im (α) there exists an N such that Im $(\alpha_n) \subseteq V$ for all $n \geq N$.

In the language of [BE] we may rephrase Exercise 4.13 by saying that $\{\alpha_n\}$ converges to α in $(C_K(p,S), \rho)$ if and only if $\{\alpha_n\}$ "converges to α in the C^0 topology on curves" (also see Section 6 of [Pen]). Corollary 2.19 and Proposition 2.21 of [BE] now combine with this observation to yield the following conclusion: Any sequence in $(C_K(p,S), \rho)$ has a convergent subsequence. The proof relies heavily on the compactness of K and on the Arzela-Ascoli Theorem from general topology. Less rigorous sketches for constructive proofs of the existence of such limit curves may be found in Theorem 6.5 of [Pen] and Lemma 6.2.1 of [HE]. We shall not give any further details here, but will simply rephrase the result by recalling that a metric space is compact if and only if each of its sequences has a convergent subsequence.

Lemma 4.5.6. If M is a (globally hyperbolic) spacetime, S is a Cauchy surface in M, p is in int $D^-(S)$ and $K = J^+(p) \cap J^-(S)$, then the metric space $(C_K(p,S), \rho)$ is compact.

Of course, the same is true if $p \in$ int $D^+(S)$ and $K = J^-(p) \cap J^+(S)$.

Having supplied $C_K(p,S)$ with a compact metric structure it remains tc define the Lorentzian length functional on $C_K(p,S)$ and hope for enough continuity to insure the existence of a maximum value. Consider then a future causal curve $\alpha : [0,1] \to M$ in $C_K(p,S)$. If α is smooth its Lorentzian length is defined as usual by

$$L(\alpha) = \int_0^1 (-g(\alpha'(t), \alpha'(t))^{1/2} \, dt \ . \qquad (26)$$

If α is not smooth its velocity vector $\alpha'(t)$ need not exist for all t and so definition (26) seems to require modification. The quickest way out of this difficulty is to appeal to a result from real analysis. Any future causal curve satisfies a local Lipschitz condition and so is differentiable almost everywhere (i.e., except on a set of Lebesgue measure zero). Consequently, the integral in (26) is defined for any such curve and we may again take this as the definition of its Lorentzian length.

Remark. If these concepts are unfamiliar to you or if you simply prefer a more constructive definition, consult Section 7 of [Pen] where $L(\alpha)$ is defined in the obvious way for causal trips (sum of the lengths of the geodesic segments) and extended to causal curves by approximating them by causal trips. In either case we have defined the length functional

$$L : C_K(p,S) \to \mathbb{R}$$

Unfortunately, it is clear that this function is *not* continuous on $C_K(p,S)$ even in the best of all possible worlds, namely, Minkowski spacetime. The reason is that a sequence $\{\alpha_n\}$ of null curves in $C_K(p,S)$ (which have zero length) can converge in $(C_K(p,S), \rho)$ to a timelike curve α (which has positive length) so that $\lim_{n\to\infty} L(\alpha_n) \neq L(\alpha)$; see Figure 4.11.

Figure 4.11

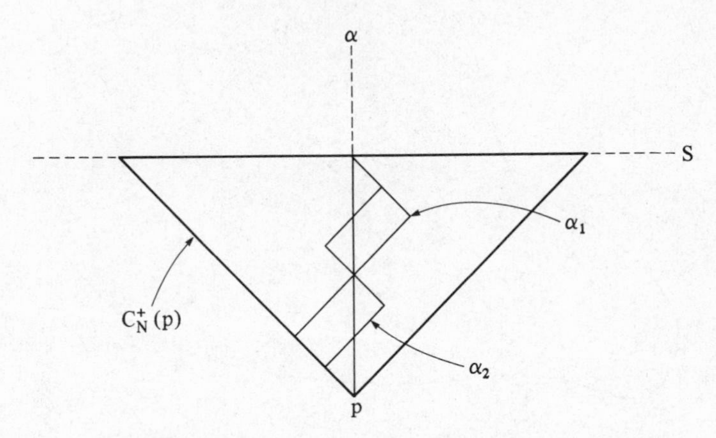

But all is not lost. Alghough, as our example shows, lim inf $L(\alpha_n)$ need not be greater than or equal to $L(\alpha)$, it is always true that lim sup $L(\alpha_n) \leq L(\alpha)$ whenever $\alpha_n \to \alpha$, i.e., we have

Lemma 4.5.7. The Lorentzian length functional $L : C_K(p,S) \to \mathbb{R}$ is upper semicontinuous on $C_K(p,S)$.

The proof of Lemma 4.5.7 depends on what seems to be a trivial observation. In Minkowski spacetime the "longest" causal trip joining two points p and q with $p \leq q$ is the straight line (geodesic). Since geodesically convex normal neighborhoods are "essentially identical to" $\mathcal{M}$ and form a base for the open sets in any spacetime it should at least be true that *locally* causal geodesics maximize the Lorentzian length. Indeed, we have

Lemma 4.5.8. Let N be a simple region in M, p and q in N and suppose the unique geodesic segment $\mu : [0,1] \to N$ from p to q is future causal. If α is any other causal trip in N from p to q, then $L(\mu) > L(\alpha)$.

The argument given in Section 7 of [Pen] for Lemma 4.5.8 is a useful reminder of how efficacious the proper choice of a coordinate system can be. One first observes that if μ is null, then the conclusion is vacuously satisfied since the inequalities (24) and (25) imply that there are no other causal trips in N from p to q. Thus, we assume that μ is timelike. Choose a point $r \in N$ to the past of p on the extension of μ.

Figure 4.12

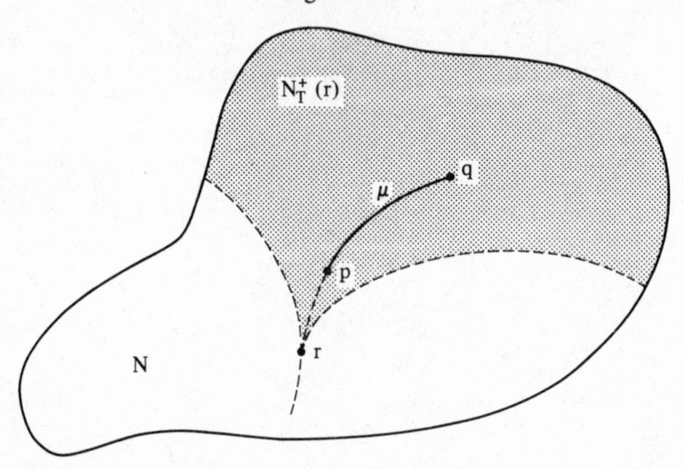

In $T_r(M)$ select an orthonormal basis $\{e_1, e_2, e_3, e_4\}$ with e_4 future timelike and let χ be the corresponding Minkowski normal coordinate patch on N. On $N_T^+(r) = \{(x^1, x^2, x^3, x^4) : x^4 > ((x^1)^2 + (x^2)^2 + (x^3)^2)^{1/2}\}$ we introduce a new coordinate patch $\bar{\chi}$ with coordinates $(\bar{x}^1, \bar{x}^2, \bar{x}^3, \bar{x}^4)$ defined by

$$\bar{x}^i = x^i / x^4, \quad i = 1,2,3$$

$$\bar{x}^4 = ((x^4)^2 - (x^1)^2 - (x^2)^2 - (x^3)^2)^{1/2} .$$

Observe that those curves of the form

$$\bar{x}^i = \bar{x}^i_0 \text{ (constant)}, \quad i = 1,2,3$$

$$\bar{x}^4 \text{ varying}$$

which lie in $N_T^+(r)$ are the timelike radial geodesics from r and the parameter $\bar{x}^4$ measures proper time along these geodesics. Thus, $g(\bar{\chi}_4, \bar{\chi}_4) = -1$. Moreover, since the Gauss

Lemma implies that these radial geodesics are orthogonal to the $\bar{x}^4$ = constant hypersurfaces, $g(\bar{\chi}_4, \bar{\chi}_i) = 0$ for $i = 1,2,3$.

Exercise 4.5.6. Describe geometrically and interpret physically the *synchronous coordinate system* $(\bar{x}^1, \bar{x}^2, \bar{x}^3, \bar{x}^4)$.

If one now calculates the metric components $\bar{g}_{ab} = g(\bar{\chi}_a, \bar{\chi}_b)$ relative to $\bar{\chi}$ the result is of the form

$$[\bar{g}_{ab}] = \begin{bmatrix} \bar{g}_{11} & \bar{g}_{12} & \bar{g}_{13} & 0 \\ \bar{g}_{21} & \bar{g}_{22} & \bar{g}_{23} & 0 \\ \bar{g}_{31} & \bar{g}_{32} & \bar{g}_{33} & 0 \\ 0 & 0 & 0 & -1 \end{bmatrix}$$

where $[\bar{g}_{ij}]_{i,j=1,2,3}$ must be positive definite since the $\bar{x}^4$ = constant hypersurfaces are spacelike. We use these metric components to calculate the length of α (using $\bar{x}^4$ as parameter and deleting the set of measure zero on which α can fail to be differentiable : If $\bar{x}_p^4$ and $\bar{x}_q^4$ are the $\bar{x}^4$-coordinates of p and q respectively, then

$$L(\alpha) = \int_{\bar{x}_p^4}^{\bar{x}_q^4} \left(-\bar{g}_{ab} \frac{d\bar{x}^a}{d\bar{x}^4} \frac{d\bar{x}^b}{d\bar{x}^4} \right)^{1/2} d\bar{x}^4$$

$$= \int_{\bar{x}_p^4}^{\bar{x}_q^4} \left(1 - \bar{g}_{ij} \frac{d\bar{x}^i}{d\bar{x}^4} \frac{d\bar{x}^j}{d\bar{x}^4} \right)^{1/2} d\bar{x}^4 .$$

Since $[\bar{g}_{ij}]$ is positive definite this will clearly be a maximum precisely when the $\bar{x}^i$, $i = 1,2,3$, are constant and this gives the geodesic μ.

With Lemma 4.5.8 in hand, 4.5.7 follows with relative ease. One considers a sequence $\{\alpha_n\}$ in $C_K(p,S)$ which converges to α in $C_K(p,S)$. It must be shown that

$$\limsup L(\alpha_n) \leq L(\alpha) . \qquad (27)$$

Toward this end we select a subsequence and relabel so that $L(\alpha_n) \to \limsup L(\alpha_n)$. Observe next that by piecing together segments, it is enough to prove our result (27) when α is a geodesic which, together with all the α_n, lie in some simple region N.

Figure 4.13

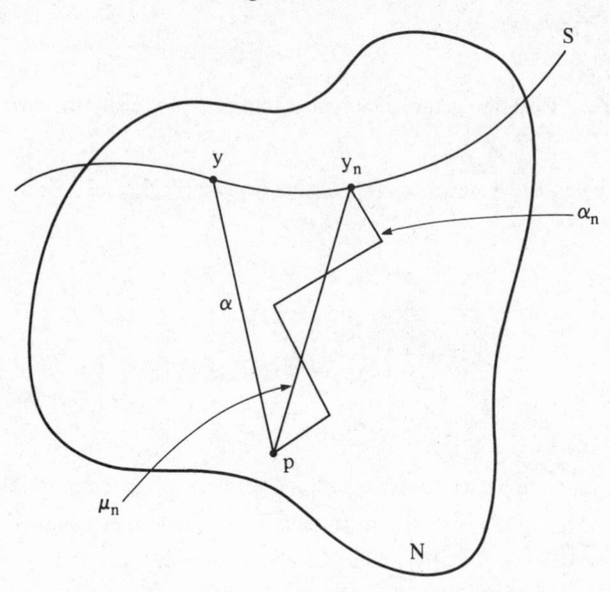

Let $y, y_1, y_2, y_3, \ldots$ be the points in S joined to p by the curves $\alpha, \alpha_1, \alpha_2, \alpha_3, \ldots$ respectively. Then $\{\alpha_n\} \to \alpha$ in $C_K(p,S)$ implies that $\{y_n\} \to y$. Let μ_n be the unique geodesic in N from p to y_n. Then

$$L(\mu_n) = (-\Omega_p(y_n))^{1/2}$$

and $L(\alpha) = (-\Omega_p(y))^{1/2}$. Since Ω_p is continuous on N, $\lim L(\mu_n) = L(\alpha)$. But, by the local maximality of geodesics (Lemma 4.5.8), $L(\alpha_n) \leq L(\mu_n)$ for every n. Thus,

$$\limsup L(\alpha_n) \leq \lim L(\mu_n) = L(\alpha)$$

as required.

Lemmas 4.5.6 and 4.5.7 now combine to give the existence of a causal curve in $C_K(p,S)$ which maximizes the Lorentzian length of all causal curves from p to S.

Theorem 4.5.9. Let M be a spacetime with Cauchy surface S and p a point in int $D^-(S)$. Then there exists a future causal curve λ in $C_K(p,S)$, where $K = J^+(p) \cap J^-(S)$, whose Lorentzian length $L(\lambda)$ is at least as large as the Lorentzian length of any other element of $C_K(p,S)$. Moreover, λ is a timelike geodesic which intersects S orthogonally.

It is only the last sentence of Theorem 4.5.9 that we have not proved. That λ must be a timelike geodesic which intersects S orthogonally is the obvious Lorentzian analogue of the fact that, in $\mathbb{R}^3$, the minimal distance from a point to a (complete, boundaryless) surface not containing the point is realized by a straight line through the point and perpendicular to the surface. The proof takes a bit of work, but the underlying ideas are clear enough from the analogy with surfaces. Being maximal, λ must be a smooth curve since otherwise the corners could be rounded off to produce a longer curve. The maximal distance from p and S cannot be achieved along a null geodesic (the length of which is zero). However, λ must be a geodesic since a maximal curve must also *locally* maximize the Lorentzian distance between any two of its (nearby) points and, by Lemma 4.5.8, this is accomplished by timelike geodesic segments. Finally, a variational argument based on an alternate version of our Theorem 4.2.3 (the First Variational Formula) shows that if λ does not meet S orthogonally it can be perturbed slightly near S to produce a longer timelike curve from p to S. For the details we refer to Proposition 11.25 of [BE].

Theorem 4.5.9 together with its obvious time reversed version ($p \in \operatorname{int} D^+(S)$, $K = J^-(p) \cap J^+(S)$, etc.) complete the proof of Theorem 4.1.1 and our work is at an end.

PROBLEMS

4.A. Geodesic Deviation, Jacobi Fields and Conjugate Points

Let $\tau : [a,b] \times (-\delta, \delta) \to M$ be a two-parameter map all of whose longitudinal curves are geodesics. Then τ is called a *one-parameter family of geodesics*. Let μ denote the geodesic $u \to \tau(u, 0)$, V the variation vector field of τ along μ, $D_{\mu'} V$ its covariant deriavtive along μ and $D^2_{\mu'} V = D_{\mu'}(D_{\mu'} V)$ its second covariant derivative along μ.

1. Show that

$$D^2_{\mu'} V = R(\mu', V)\,\mu' \; . \tag{28}$$

If μ alone is given, (28) is a differential equation, called the *Jacobi equation* or *equation of geodesic deviation*, the solutions V of which are called *Jacobi fields* along μ. If the longitudinal curves are all timelike, then (28) can be regarded as an analogue of Newton's Second Law in that it describes the relative accelerations of nearby freely falling material particles ($D^2_{\mu'} V$) in terms of the geometry of the gravitational field (R).

2. Show that a Jacobi field along μ is uniquely determined by prescribing V and $D_{\mu'} V$ at some point.

3. Show that if $\mu : [a,b] \rightarrow M$ is a geodesic and V is a Jacobi field along μ, then $g(V(t), \mu'(t)) = mt + b$ for some constants m and b. *Hint.* Differentiate $g(V(t), \mu'(t))$ twice with respect to t and use (28).

4. Show that if the Jacobi field V along the geodesic μ vanishes at two distinct values of t in $[a,b]$, then V is everywhere orthogonal to μ'.

   Two points p and q on the geodesic μ are said to be *conjugate along* μ if there is a nontrivial Jacobi field along μ which vanishes at p and q. Intuitively, the existence of conjugate points along μ indicates that some family of nearby geodesics can "cross" at these points. Such behavior has important geometrical and physical implications (see Section 7 of [Pen]). It is, however, strictly a global phenomenon:

5. Show that if μ lies inside a simple region N, then a Jacobi field along μ is uniquely determined by its values at any two points. Conclude that μ has no conjugate points in N.

4.B. Strongly Causal Spacetimes

An open set U in the spacetime M is *causally convex* if and only if for all p and q in $U, p \ll r \ll q$ implies $r \in U$.

1. Give a number of examples in $\mathcal{M}$ of open sets that are (are not) causally convex. In particular, find a local base at any p in $\mathcal{M}$ consisting of open sets that are (are not) causally convex.

   M is *strongly causal* if each p in M has a local base of neighborhoods which are all causally convex.

2. Show that a strongly causal spacetime satisfies the chronology condition.

3. Show that M is strongly causal if and only if every point in M has a local base of neighborhoods of the form $I^+(p) \cap I^-(q)$, where $p \ll q$.

4. Show that a stably causal spacetime is strongly causal.

   All of this and more can be found in Section 4 of [Pen].

4.C. Distance: Riemannian and Lorentzian

If M is a manifold with metric g (either Riemannian or Lorentzian) and $\alpha : [a,b] \rightarrow M$ is a smooth curve, then the length of α is defined as usual by

$$L(\alpha) = \int_a^b |g(\alpha'(t), \alpha'(t))|^{1/2} \, dt \ .$$

The length of a piecewise smooth curve is defined by adding the lengths of its smooth segments. Let p and q be two points of M. Using the connectedness of M one can show that p and q can always be joined by some smooth curve in M (see [BG]). Now, if g is Riemannian we define the *Riemannian distance* $d_R(p,q)$ *from* p *to* q by $d_R(p,q) = \inf \{L(\gamma) : \gamma \text{ is a piecewise smooth curve in } M \text{ from } p \text{ to } q\}$.

1. Show that $d_R(p,q) \geq 0$, $d_R(p,p) = 0$, $d_R(q,p) = d_R(p,q)$ and $d_R(p,q) \leq d_R(p,r) +$ $+ d_R(r,q)$ for all p, q and r in M.

   It is also true, although more difficult to prove, that $d_R(p,q) = 0$ implies $p = q$ so that d_R is a "metric" in the sense of point-set topology. Moreover, a subset of M is open in M if and only if it is open in the metric space (M, d_R). See Section 9, Volume 1, of [Sp2].

   If M is Lorentzian we define the *Lorentzian distance* $d_L(p,q)$ from p to q by setting $d_L(p,q) = 0$ if $q \notin J^+(p)$ and, if $q \in J^+(p)$, $d_L(p,q) = \sup \{L(\gamma) : \gamma \text{ is a piecewise smooth causal curve (or causal trip) from } p \text{ to } q\}$.

2. Show that if M does not satisfy the chronology condition, then there exists a p in M such that $d_L(p,p) = \infty$.

3. Show that $d_L(q,p)$ is generally not equal to $d_L(p,q)$.

4. Show that, for all p, q and r in M, $d_L(p,q) \geq d_L(p,r) + d_L(r,q)$.

   We must conclude then that d_L is nothing at all like a "metric". It is, nevertheless, a very useful device. Indeed, [BE] bases much of its study of Lorentzian geometry on properties of d_L.

4.D. Isometries

Let M and $\overline{M}$ be manifolds with metircs g and $\overline{g}$ (both Riemannian or both Lorentzian) and let $f : M \to \overline{M}$ be a diffeomorphism of M onto $\overline{M}$. At each point p in M, f induces an isomorphism $f_{*p} : T_p(M) \to T_{f(p)}(\overline{M})$ (see Problem 3.E). For each $\overline{p}$ in $\overline{M}$ we let $p = f^{-1}(\overline{p})$ and define a real valued function denoted f^*g on $\bigcup_{\overline{p} \in \overline{M}} T_{\overline{p}}(\overline{M}) \times T_{\overline{p}}(\overline{M})$ by

$$f^*g(\overline{v}, \overline{w}) = g(f_{*p}^{-1}(\overline{v}),\ f_{*p}^{-1}(\overline{w}))$$

for all $\bar{v}$ and $\bar{w}$ in $T_{\bar{p}}(\overline{M})$.

1. Show that $f^* g$ is a metric for $\overline{M}$ of the same type as g (and $\bar{g}$). $f^* g$ is called the *metric on* $\overline{M}$ *induced by f.*

   The map f is called an *isometry* if $f^* g = \bar{g}$, i.e., if it preserves inner products in the sense that

$$g(v,w) = \bar{g}(f_{*p}(v), f_{*p}(w))$$

   for all p in M and all v and w in $T_p(M)$.

2. Show that an isometry $f: M \to \overline{M}$ preserves lengths of curves and therefore preserves distance (see Problem 4.C). In the Riemannian case the converse is also true, i.e., a distance preserving map is an isometry (see Volume 3 of [Sp2]). In the Lorentzian case, however, a distance preserving map need not even be continuous. One can show, however, that a distance preserving map of a strongly causal space-time onto itself is an isometry (see [BE]).

3. Let f and F be two isometries of M onto itself. Show that if there exists a p in M at which $f(p) = F(p)$ and $f_{*p} = F_{*p}$, then $f = F$ everywhere on M.

REFERENCES

[ABS] Adler, R., M. Bazin and M. Schiffer, *Introduction to General Relativity*, McGraw-Hill, New York, 1965.

[BE] Beem, J.K. and P.E. Ehrlich, *Global Lorentzian Geometry*, Marcel Dekker, New York, 1981.

[BG] Bishop, R.L. and S.I. Goldberg, *Tensor Analysis on Manifolds*, Dover, New York, 1980.

[BoG] Bondi, H. and T. Gold, "The steady-state theory of the expanding universe", *Mon. Not. R. Astron. Soc.*, **108**, 252-270, 1948.

[CE] Cheeger, J. and D.G. Ebin, *Comparison Theorems in Riemannian Geometry*, North Holland, Amsterdam, 1975.

[E1] Einstein, A., H.A. Lorentz, H. Minkowski, etc., *The Principle of Relativity*, Dover, New York, 1958.

[E2] Einstein, A., *The Meaning of Relativity*, Princeton Univ. Press, Princeton, N.J., 1956.

[Fa] Faber, Richard L., *Differential Geometry and Relativity Theory: An Introduction*, Marcel Dekker, New York, 1983.

[G] Göbel, R., Physikal Teil (II) der Habilitationsschrift, Würzburg, 1973.

[Har] Hartman, Philip, *Ordinary Differential Equations*, Hartman, Baltimore, 1973.

[H] Hawking, S.W., "The occurrence of singularities in cosmology", *Proc. R. Soc. London* **A 294**, 511-521, 1966.

[HE] Hawking, S.W. and G.F.R. Ellis, *The Large Scale Structure of Space-Time*, Cambridge University Press, Cambridge, England, 1973.

[HKM] Hawking, S.W. and A.R. King and P.J. McCarthy, "A new topology for curved space-time which incorporates the causal, differential and conformal structures", *J. Math. Phys.*, **17**, 174-181, 1976.

[Hi] Hicks, N.J., *Notes on Differential Geometry*, Van Nostrand, Princeton, N.J., 1965.

[Ho] Hoyle, F., "A new model for the expanding universe", *Mon. Not. R. Astron. Soc.* **108**, 372-382, 1948.

[MTW] Misner, C.W., K.S. Thorne and J.A. Wheeler, *Gravitation*, W.H. Freeman, San Francisco, 1973.

[Nab] Naber, G.L., *Topological Methods in Euclidean Spaces*, Cambridge University Press, Cambridge, England, 1980.

[Nai] Naimark, M.A., *Linear Representations of the Lorentz Group*, Pergamon, New York, 1964.

[O1] O'Neill, B., *Elementary Differential Geometry*, Academic Press, New York, 1966.

[O2] O'Neill, B., *Semi-Riemannian Geometry*, Academic Press, New York, 1983.

[Pa] Pais, A., *Subtle is the Lord* ..., Oxford University Press, Oxford, England, 1983.

[Pen] Penrose, R., *Techniques of Differential Topology in Relativity*, S.I.A.M., Philadelphia, 1972.

[Per] Perrin, R., "The twin paradox: A complete treatment from the point of view of each twin", *Am. J. Phys.*, **47**, 317-319, 1979.

[S1] Spivak, M., *Calculus on Manifolds*, W.A. Benjamin, New York, 1965.

[S2] Spivak, M., *A Comprehensive Introduction to Differential Geometry*, Vol. I-V, Publish or Perish.

[SW1] Sachs, R.K. and H. Wu, General Relativity and Cosmology, *Bull. Amer. Math. Soc.*, **83**, 1101-1164, 1977.

[SW2] Sachs, R.K. and H. Wu, *General Relativity for Mathematicians*, Springer-Verlag, New York, 1977.

[Sy1] Synge, J.L., *Relativity*: *The Special Theory*, North Holland, Amsterdam, 1972.

[Sy2] Synge, J.L., *Relativity*: *The General Theory*, North Holland, Amsterdam, 1960.

[TW] Taylor, E.F. and J.A. Wheeler, *Spacetime Physics*, W.H. Freeman, San Francisco, 1963.

[W] Weyl, H., *Space-Time-Matter*, Dover, New York, 1952.

[Z1] Zeeman, E.C., "Causality implies the Lorentz group", *J. Math. Phys.*, **5** 490-493, 1964.

[Z2] Zeeman, E.C., "The topology of Minkowski space", *Topology*, **6** 161-170, 1967.

INDEX